Sleuthing
the Divine

Sleuthing the Divine

THE NEXUS OF SCIENCE AND SPIRIT

Kevin Sharpe

FORTRESS PRESS
MINNEAPOLIS

SLEUTHING THE DIVINE
The Nexus of Science and Spirit

Excerpts marked JB are from The Jerusalem Bible, copyright © 1966 by Darton, Longman and Todd, Ltd., and Doubleday and Company, Inc. Used by permission of the publisher. Scripture passages marked NEB are from The New English Bible, copyright © 1961, 1970 by the Delegates of the Oxford University Press and the Syndics of the Cambridge University Press. Reprinted by permission. Scripture passages marked NTP are from *The New Testament and Psalms: An Inclusive Version* copyright ©1995 Oxford University Press. Used by permission.

"One of Us," words and music by Eric Bazilian. Copyright © 1995, 1996 Human Boy Music (ASCAP). All rights administered by WB Music Corp. All rights reserved. Used by permission of Warner Bros. Publications U.S. Inc., Miami, FL 33014.

Cover photography copyright © 2000 PhotoDisc. Used by permission.
Book design by Timothy W. Larson

Library of Congress Cataloging-in-Publication Data

Sharpe, Kevin, –
    Sleuthing the divine: the nexus of science and spirit / Kevin Sharpe.
        p.   cm.
    Includes bibliographical references and index.
    ISBN 0-8006-3236-2 (alk. paper)
        1. Religion and science. I. Title.
BL240.2.S537 2000
215—dc2   00-037116

The paper used in this publication meets the minimum requirements of American National Standard for Information Sciences—Permanence of Paper for Printed Library Materials, ANSI Z329.48-1984.

Manufactured in the U.S.A.   AF 1-3236
04   03   02   01   00   1   2   3   4   5   6   7   8   9   10

To Mary Catherine Lacombe

CONTENTS

Preface   ix

Part One: Searching Meaning   1

1. Scientific and Spiritual Thought: Their Mutual Relevance   3
2. Connection   14
3. Separation with Connection   23

Part Two: Searching the Divine   31

4. The Subuniverse Divine   33
5. It Just Is   41
6. The Divine Acts?   47
7. What Does the Divine Do?   57
8. The Real Divine   66
9. Mystery   72
10. The Character of the Divine   81

Part Three: Searching Morality   93

11. Freedom   95
12. Liberation and Values   101
13. Histories of the Universe   106
14. Sociobiology   115
15. Does Morality Come from the Divine?   125
16. Does Morality Come from Biology?   130
17. Creating a Morality   137

Part Four: The Spiritual Quest 145

  18. Suffering and Evil and Humanization 147
  19. Christian Belief 154
  20. Scientific and Spiritual Thought: Open to Change 163

Notes 168
Bibliography 172
Index 179

A T THE CENTER OF MUCH RELIGIOUS LANGUAGE lies the word *God*. Yet that central word carries many and often conflicting meanings. Many of us pass through a religious crisis at some point in our lives, and we easily transfer our conflict to *God*. Confusion over the word *God* becomes worse when a religion thinks itself better than the others; some claim a monopoly on the word—We have truth!—as though they compete for the true God. Think about the following examples; which most correctly uses the word *God*?

- "Oh, God! I've poured bleach over the colored clothes!"
- "'God' stands for the way in which the upper classes control and hold their power. They use it to make the populace tremble in fear."
- "Oh how I love you, God. You've come into my soul and filled me with joy!"
- "I believe in God Almighty, Creator of heaven and earth, and in Jesus Christ his only begotten Son. . . ."

How might someone judge the correct usage? Does a standard exist against which to check it? Who or what is God anyway? What's God really like? Where might a person find God? Why do so many people claim to deliver God? And why do people create pictures of God? Lots of God images flutter about with no hard and fast way to decide between them. Somehow God and the images of God have become separated.

Because of that confusion and because people differ (often significantly) in what they mean by the G-word, I'm going to stop using it right here. With my decision, I stop clarifying what I *don't* mean. I look to a fresh start to help answer questions such as those above.

When the meaning of a word becomes too restricted, people often coin new words or resurrect old ones that do not have the problems associated with the word they replace. I'd rather not surrender the G-word altogether, though. It has too much significance in the lives of people I know; it has been too important in human history and experience to throw away at my whim. Instead, I simply want to look at the "divine being," God, without the limits set by religious traditions, bias, culture, or common parlance.

So I prefer to use a synonym for it. To be honest, I have yet to find a good alternative to the G-word, as all similar words bring excess meanings, but I opt for the term *Divine* as the best I can ferret out at present. It too invokes problems—think of "Darling, you look divine"—but not as many as the G-word; it has fewer associations. This advantage also produces its weakness, however: the G-word holds many meanings without much explanation, and in using *Divine*, I lose that strength.

I replace other words, too. Tradition would label my work *theology*, but I avoid that word, as it also carries too much baggage. Instead, I employ terms such as *spiritual thinking* and *spiritual ideas* and *a system of spiritual ideas*. Similarly, I use synonyms for words such as *religion*—but not for *spirituality*. *Spirituality* escapes my habit to replace words, because it carries its own proper and frequent use. It has a positive meaning centered on a way of life, which I want to leave untarnished by association with my use of *spiritual*.

I feel strongly about using alternatives. I write not only for those in an orthodox fold who are interested in other ways to look at the Divine, but also for those interested in spiritual matters and those who have left a fold to search for meaning elsewhere. The words I use try to include them. The name *Divine* confuses less than *God* and so should turn off fewer people.

The task I set for myself is to reconstruct the word *Divine* so it attains significance. This is difficult to do. I can't salvage the word with an instant system of spiritual ideas. Many thinkers, myself included, have wrestled long and hard with this problem—and the solutions, I believe, remain sterile. So I start again.

After much reflection, I start this reconstruction from two points. First, I follow an assumption in trying to understand the Divine: the Divine is real. The Divine exists. But secularization and the rise of modern science have gradually changed the way people think and feel. They have eroded the belief systems of medieval and earlier Christendom, including people's sense of reality about that divinity. Whatever picture I come up with, then, the divinity it depicts must appear credible and real for modern, secularized humanity.

Why do I want to picture the Divine as real? Because, to start with, this is a basic property of the Divine, of any properly functioning divinity. Only a dishonest dealer would sell a car "in good working order" when it didn't have an engine. The Divine needs at least the "engine" of reality. Believers accept the reality of their divinity, and thus I make reality a must for what I construe. I shy away from fantasy.

When people ask me if I believe in the Divine, I reply yes without hesitation. I then ask, "What divinity?" I have yet to endow the word with content. My belief in the Divine's reality leaves open what I mean by *the Divine;* it stops short of explaining what the Divine is. To crave an image that portrays a real divinity assumes nothing further about the Divine.

The second starting point for the new picture of the Divine focuses on the word *secular.* We live in the modern Western world with all its corruption and inhumanity, idealism and hope. And many of us relish it: the universe, our society, and our lives feel good. Most Westerners place the centers of their realities in the secular present, in that which we describe as "here." We live as secular people; our spiritual ideas must build from the secular "here." Here is exciting and challenging. The answer to inhumanity and to modernity's problems arises, we believe, from the secular here and not from elsewhere. I therefore resist focusing on another world, spiritual or otherwise. The roots of the Divine extend deep into the secular.

With these simple assumptions, and a prying inquisitiveness, I invite you to join me in sleuthing the Divine.

Part One
Searching Meaning

SCIENTIFIC AND SPIRITUAL THOUGHT
THEIR MUTUAL RELEVANCE

Reconstructing the Divine

MANY BELIEVERS TREAT SCIENTIfiC THOUGHT AND SPIRITUAL THOUGHT as mutually exclusive. A spiritual movement that seeks to free oppressed people may reject technology and science because it sees them used to exploit and deprive dominated people of basic human rights. A local school board made up of fundamentalists and arch-conservatives who dismiss science because it contradicts what they believe may demand the teaching of creationism and the ignoring of evolution. Charismatics and Pentecostals immerse themselves in experience of their spiritual realm and discard the scientific-secular approach as connecting with the real.

Which holds authority and truth in the opinion of the spiritually inclined: the approach of the secular and scientific world, or the spiritual, the church? The above three caricatures—liberationists, fundamentalists, and charismatics—avoid the contest because their beliefs mandate their claim to authority and truth. Some say the spiritual should rightfully hold the position of authority. They partition the secular-scientific off from the spiritual. This book, however, promotes a more neutral position—perhaps leaning toward the secular-scientific—while retaining the integrity of the spiritual. And this more neutral position requires a reworked idea of the Divine. We need a new, more adequate image that depicts the Divine as real and essential. The reconstructed Divine must act in the universe in ways that strike us as obvious. And the new model must root itself in the secular and scientific.

What do most people accept as real? Reality for the majority of us concerns consumer goods and other people in the context of the secular

world. We bestow great importance on the secular-scientific experience
and what it produces. Most people, therefore, would feel the reality of the
Divine if a close relationship existed between secular-based science and our
understanding of the Divine, because the feeling of reality would transfer
from secular technology-science to the Divine. A close relationship
between the two would produce, for most of us, a real divinity.

To move the inference in the opposite direction, start with a real divinity. You know the reality of the chair you sit in because it holds you up off
the floor and its arms support your elbows. The feeling of reality comes
from your interactions with it. Similarly, the reality of the interactions
between the Divine and the universe leads to the reality of the Divine.

The reality of the chair also involves gravity, because otherwise it
could float off in one direction and you in another. Sophisticated scientific
theories thus play into your feeling its reality. Sophisticated scientific theories also play into our feeling the reality of the Divine; our understandings
of the Divine's interactions with the universe relate to our understandings
of how the universe operates. And the more real we find the Divine's interactions, the more frequently we must relate our understanding of the
Divine to our understanding of the universe. A real divinity leads to a close
relationship between spiritual thought and science.

Establishing Mutual Relevance

These two subjects—the relationship between spiritual and scientific
thought, and the reality of the Divine—emerge as much the same. They are
equivalent. Their relationship represents the challenge of secular modernity
to spiritual traditions, the most important challenge in a long time. The
challenge also brings into question the status of the Divine. Many, if not
all, of the concerns about the Divine also show up in the intersection of science and spiritual thought. The criterion—that the relationship between
scientific and spiritual thought must evince *mutual relevance*—can thus
help ensure the Divine's reality.

Their relevance has not always been affirmed. In 1903, British
philosopher and mathematician Bertrand Russell epitomized a mood:

> Science can help us to get over this craven fear in which [humanity] has
> lived for so many generations. Science can teach us, and I think our own

heart can teach us, no longer to look around but rather to look to our own efforts here below to make this world a fit place to live in, instead of the sort of place that the churches in all these centuries have made it.[1]

In science's dominance, the attitude that the spiritual offered nothing to science extended further for Russell: a respectable philosophical agenda included only negative things about anything spiritual. This torpedoed toward mutual relevance. But opinions change. A book from William Austin, a philosopher at the University of Houston, reflects the easing out of that harsh mentality and the ushering in of greater openness to spiritual matters. Published in 1976, *The Relevance of Natural Science to Theology* justifies its title's claim by refuting opposing ideas. The tool Austin applies is the word *relevance,* and I develop my meaning for the word from his.

Do all people descend from Adam and Eve, whom Yahweh fashioned "of dust from the soil" (Gen. 2:7 JB)? Did the begetting of one couple, who had no parents, lead to every other human being? Biologists think not; biology insists on parents for everybody. In Austin's terms, biology has "direct relevance" to the Genesis story. A collection of statements in one field bears direct relevance to a statement in another field, he writes, if we can deduce it or its opposite from the collection. In the Genesis case, the opposite of what biology says can lead to the spiritual understanding of Eve and Adam.

William Paley (1743–1805) believed that the Divine engineered the members of a species to function as they do. The parts of the universe work together in a neat and orderly fashion because a neat and orderly mind produced them. Charles Darwin (1809–1882) offered an alternative explanation for the way organisms adapt to their environments: only those inherently fit to survive an environment live and pass these characteristics on to their offspring. Austin cites these two approaches as an example of "quasi-direct relevance," applying the definition to two theories—from different fields or the same one—if they offer alternative, apparently competing explanations of the same data.

Returning to human evolution and the Eve and Adam stories, note that the two theories cover some of the same data—for instance, the remnants of several early cultures. Note, second, that evolutionists insist on parents for everyone and that this forces the theories to compete. These two points establish what I call *relevance,* an idea that builds on Austin's direct and quasi-direct forms. Relevance happens if, first, the fields cover some of the same data. Then, second, either the first field includes statements directly relevant for statements in the second, or the two fields include theories with quasi-direct relevance for each other. Thus, science has relevance for spiritual thought over the origin of the human species.

Not only is science relevant to spiritual thought, as in the Genesis example; spiritual ideas are, in some cases, relevant to science. When this happens, scientific and spiritual ideas acquire mutual relevance. They may then compete with each other; though they may ask different questions, their answers overlap in places. In principle, they can then exchange ideas or methods, recognize and resolve conflicts, and support the other's ideas. The Genesis and biological accounts of the origin of the human species possess this mutual relevance because the relevance extends both ways. We wait to see, though, if they eventually resolve their clashes.

Failure to identify the mutual relevance of the two kinds of thinking leads to truncated understanding in each realm. For example, the Chernobyl nuclear power plant in the Ukraine entered world history on April 26, 1986. An improperly supervised experiment led to an uncontrolled reaction. The plant's operators had turned off the water-cooling system, and the buildup of steam caused an explosion, blowing off the reactor's protective covering. The release of radiation into the air spread to Russia and Europe; many people in Russia died and many more are still dying. Practiced and applied science needs humanizing. Similarly with the spiritual: the absolute decree against birth control speaks to the way spiritual leaders can require religious practices over human suffering, let alone over scientific and technological expertise. Further, much of the dehumanizing that crops up in the world comes from the split between scientific and spiritual thought, part of the larger division between fact and value. For instance, technology and science give birth to the World Wide Web, but no ethics appear with it and child pornography rushes in. The power and powerlessness of both the spiritual and scientific traditions for good and for ill touch everyone. Sighting the mutual relevance between the two traditions could help remedy the oversights of each.

Scholars can put this relevance into effect by developing a science and a system of spiritual thought that clearly show that relevance. Consider what this requires. Few people question one direction of the relevance, that of science for spiritual thought. Even many who support a literal interpretation of the creation accounts in Genesis, for instance, try to fit their ideas with evolutionary theory. "How long is a day in the life of the divine creator?" they frequently ask. "How long did each stage, including human life, last in the evolution of the universe? A thousand years? Ten thousand? A million?" That's one direction of the relevance.

An evangelist once insisted that the Divine kept a hurricane from hitting a town because many believers lived there. The weather service may want to chew over its competing explanation for the hurricane's change of course. When spiritual thinkers claim that the Divine acts or intervenes in the universe of natural causes, they refer to nonspiritual events and so touch a scientific discipline.

That people undergo spiritual experiences falls under the scrutiny of psychology and psychiatry; maybe scientists can apply brain functioning to explain these events. University of Pennsylvania neuroscientists Eugene d'Aquili and Andrew Newberg have described a neuropsychological model to help us understand mystical happenings, from a simple sense of awe to the sublime states of altered consciousness that follow years of meditation. They isolate the segments of the brain that yield spiritual, mystical, ritual, and mythic experiences. But the question remains: as my experience of the chair I sit on connects with a real chair outside my brain, does a spiritual experience connect with a real divinity, or is it something the brain concocts? Here, spiritual claims engage scientific theory.

Engaging the Spiritual and the Scientific

Many spiritual beliefs enter the universe that science attempts to explain, but science writes off hurricane-like examples of supposed divine action. It's difficult, though possible—as we will see later—to find instances where spiritual claims more critically influence scientific theory.

Sometimes the discussion between science and spiritual thought brings a smile, as with the evangelist and the hurricane. Sometimes the mass media even treats it seriously and with respect. An example, from journalist Robert Wright, appeared as the cover story of a Christmas issue of *Time* magazine and represents the best of current research on the subject. Does Wright's work show mutual relevance?

Wright notes that the Divine once had charge of the beginning and end points of time, and we had no right to debate them. Now Cambridge

physicist Stephen Hawking thinks he can decide what happened at the
beginning. His theories typify how science pushes ahead: the sweep of its
explanations advances relentlessly, even invading turf that past generations
held sacred. Wright supplies more examples:

> The most ethereal parts of life—the things that once seemed heaven-
> sent—have fallen steadily within reach of concrete explanation. The
> mapping of our finer feelings to neurotransmitters and other chemicals
> proceeds apace. Love itself—the love of mother for child, husband for
> wife, sibling for sibling—may boil down, in large part, to a chemical
> called oxytocin. . . . Another bit of less-than-inspiring news is the clearer,
> more cynical, understanding of why love exists—how it was designed by
> evolution for only one discernible purpose: to spread the genes of the
> person doing the loving.[2]

Wright suggests that modern science dampens hopes for a supra-
human meaning to life and the universe. It removes the inexplicable. But,
he continues, something else also emerges from these developments. The
closing stages of twentieth-century science provided fertile ground for
bona fide spiritual speculation. Many scientists, Wright points out, step
back to view a bigger picture than science describes and conclude that an
intelligent designer produced it. The Divine lies behind how the bits of the
universe fit together, unfolding a purpose that includes human experience.
We are meant to be here. With this conclusion arrives the relevance of science for spiritual thought.

The anthropic principle starts with the fact that the universe has certain physical constants, such as the speed of light. A slight variation in the
value of any one of them from what it is would render the universe unsuit-
able for the evolution of creatures like ourselves. Wright offers two ways to
explain or understand this, two types of the anthropic principle. One says
that a huge number of universes exists and ours just happens to inherit
these values for the constants, which is why humans exist here in this universe. The other says that a divine creator designed the universe and its
constants this way because "the plan" called for humanlike creatures.

Like many other inquirers, Wright finds it improbable that a living
molecule would form randomly in a primeval soup. A new field of science,
named complexity theory, studies the emergence and development of life
and promotes self-organization as the way to understand it. Facing
increased disturbances, says the theory, physical systems can become more

structured. Even mundane systems such as water and air can demonstrate this. Think about it the next time a tap releases a little bit of water that runs around the lip before gathering itself into a regular drip. The Belgian chemist Ilya Prigogine, winner of a Nobel Prize, helped pioneer this way of thinking. Take a physical system pushed out of a stable state. It may become stable again, Prigogine calculates, but at a level of organization higher than the original. In such a way, self-organization may account for the origin of life, giving it a better chance of happening than if left to the roulette wheel. Maybe self-organization is a new fundamental law of science.

The minute chance of life forming randomly inclines many spiritual thinkers to believe the Divine intervened on the earth to create it. It could not happen by itself, they argue. The new science of complexity with its suggested law of self-organization removes this role for the Divine. But it still leaves a task—though a different one than thought before—for the Divine. The pressure to self-organize suggests a divinity that slants the universe to produce life, a divinity that gives the universe a special law to increase complexity. Divine design would produce a universe more inclined to create life, Wright realizes, than one where life happens by chance. Science once again unfolds divine purpose.

What about the origin of *intelligent* life, though? Wright thinks this too is inevitable. The same species that lives in the frigid arctic also lives in the boiling tropics, in the airlessness of space and in the pressurized depths of oceans; human beings show a lot of flexibility. Natural selection favors flexible behavior and the intelligence that frequently produces it because such adaptations help reproduction and survival. The genes that bestow it flourish—and humans overpopulate the planet. Princetonian John Tyler Bonner follows this line of reasoning to show that natural selection can by itself produce more and more intricate organisms. If Bonner is correct, Wright suggests, natural selection produced the human brain and intelligence.

From one point of view, this account of human life flows from physical ideas. From another, given the laws and tendencies in the universe toward complexity, life, and intelligence, it rests on a notion of the Divine. The Divine biased the universe to work this way so it would produce human beings.

When it comes to humans, though, there is still the need to account for purpose and emotions. Meaning and the moral qualities of life arise from feelings, Wright says. As well as responding to the universe, humans experience it. How did feelings emerge? Oxford University don William Hamilton suggests kin selection as an answer. People feel altruism to their kin and act on it. This behavior evolved because the genes responsible for it lie not only in those who exhibit it but also in their relatives. An act of altruism, even if it means the death of the altruist, saves those genes. Altruism aids the survival of a kin group, so the chromosome units that promote it continue to reproduce and increase in influence. These successful genes provide the feeling of altruism.

Kin selection is a force in the evolutionary story that led from intelligent life to the human feelings of compassion, affection, and love. A gambler at the big bang could have laid good odds on the appearance of this type of love. Again, Wright plugs for the larger purpose of this in the scheme of things.

We should applaud Wright for raising the challenges of science to traditional spiritual ideas. The Divine doesn't intervene in the universe to spearhead such human experiences as the sense of morality, Wright says. Science, rather, describes the Divine at work; the Divine produced these abilities through natural laws.

Considering Purpose

For Wright, spiritual thinkers should discuss the purpose of the universe that science describes, an attitude that nearly all writers in the area of science and spiritual thought assume. Harvard astrophysicist Owen Gingerich says the choice really lies between "purpose and accident."[3] Did the universe with its humans come about because the Divine built its likelihood into the laws of nature? Or did it arise purely by chance? This is a rhetorical question for Gingerich. Purpose enters where science ends and the search for spiritual faith starts. Purpose is the key to the universe. The Divine had reasons to create it, science hints at them, and spiritual thinkers point them out. The mutual relevance of scientific and spiritual thought starts to emerge, and the discussion continues where Wright stops, into the implications of science for qualities such as purpose.

Since we have them, should the Divine possess two hands, each with five fingers? Does the Divine crave food and sex, and does the chemical oxytocin flow through divine veins when the Divine loves? Of course not. The

Divine isn't a biological being that evolved under environmental pressures and so doesn't have hands and fingers, the need for food and sex, or veins throbbing with oxytocin. Similarly, purpose, love, and other such qualities evolved biologically; they pertain to organic creatures that developed under the imperative to survive genetically. An inherited trait such as the need for purpose features specifically in animals such as humans. So why should the Divine have purposes, in particular, a purpose for the universe? Purpose belongs with biochemicals and genes; purpose isn't the key to the universe.

How do spiritual thinkers react to this? One response asserts the irrelevance of this science to spiritual thought; another claims the science will collapse under scrutiny; another emphasizes the tentative nature of all scientific knowledge. But what if science supports it and mutual relevance applies? Some say the Divine does hold purposes and employs evolution to install this attribute into humans. The pressures of natural selection involve many ups and downs, however. To control each and every one of them, the Divine must constantly intervene in the universe. This opposes the flow of Wright's argument that the Divine originally loaded the creation with human-producing laws and intervention plays no part. Evolution also requires genes to change, a quantum-physical, purely-by-chance phenomenon. Many spiritual thinkers consider such events to lie outside even the Divine's power to predict; thus, the Divine couldn't imagine what evolution would produce, unless, again, massive intervention occurred. It is highly unlikely, therefore, that the Divine set up evolution to reproduce purpose in humans.

Though the science is new, such spiritual positions follow a long-standing historical precedent. What happened after the church's conflict with Galileo? The belief that the earth occupies the center of the universe slipped away. With Darwin, the Genesis details of creation quietly left. With Freud, the mind's connection to the Divine said good-bye. Spiritual thought frequently conflicts with new scientific theories. In time, spiritual thinkers adjust to the developments, usually by redefining their subject far from the current science. An emphasis on love and purpose developed from such stances when the physical, biological, and mental worlds withdrew from what spiritual thinkers might respectably explain. Now we question even love and purpose as divine attributes and bases for spiritual thought. The response of spiritual thinkers follows the historical pattern.

A common strategy for a spiritual defense against science's claims affirms that as humans are free to follow their desires, so the universe operates independently of the Divine. The universe freely follows its own laws set up by the Divine at creation. The Divine and the universe are separate. Then how does the Divine act in the universe? The divine Spirit moves a woman to speak in foreign tongues. A man's liver cancer disappears overnight after a prayer meeting. The Divine, tradition continues, intervenes in this independent universe by acting in specific, miraculous, and mysterious ways. So the Divine steps in where science steps out. Likewise, the Divine steps out where science steps in. The Divine doesn't distribute justice for wrong behavior through thunderstorms and lightning or illness. Science forces believers to stop thinking of the weather and disease as divine interventions. As science progresses, the Divine performs less and less. So it is with purpose. The Divine didn't set up the evolutionary process to produce beings that think and feel in heavenly and hellish ways. Rather, in their attempts to understand, people forged the Divine's nature from these human terms and capacities. Science limits this practice; the more it advances, the fewer ways remain for the Divine to act and be like a human.

Acknowledging the Challenge

Yet this model of how the Divine operates and how scientific and spiritual thought relate—imputing totally separate spheres presided over by an utterly transcendent divinity that intervenes for special occasions and purposes—faces severe problems. The belief that the Divine exists beyond the universe aggravates the problem; the further away the Divine, the harder it is to understand how the Divine interacts with the universe. As the traditional beliefs slip away, so does the traditional divinity. Any idea of the Divine that survives the light of science, any idea of the Divine that emerges· from mutual relevance, must avoid the Divine's "total beyondness."

Other problems arise when spiritual thinkers overemphasize the difference between the spiritual and scientific worlds or when they avoid the difficult questions. Doing so hampers a free and full flow between their ideas and those of science. When they play down the relevance of the spiritual for scientific culture and inhibit the spiritual's ability to speak to the secular, they block the spiritual assumption to address all aspects of life.

Less and less does the spiritual then provide an authoritative moral direction for science and technology; fact and value move farther apart.

A better way to meet the challenge is to emphasize the relevance of scientific and spiritual ideas for one another. This means actively exploring their points of contact. It means working out a flexible system of spiritual ideas that moves with scientific advances and builds on the findings of science, adopting a method like science's. It means exploring scientifically. It means that spiritual thought offers hypotheses and insights for scientific scrutiny. And it means promoting a science that seeks advances in spiritual thought, motivating the search for scientific explanations that fit with spiritual belief. Spiritual thinkers and scientists could then see themselves working in tandem with each other.

This is difficult. To resolve specific conflicts in ways that uphold mutual relevance is serious business. It may lead to changes in the initial theories, even abandoning one or both of the troublemakers. To sort through the confusion that arises from assigning human qualities to the Divine requires us to act this out. Spiritual thinkers must clarify the nature of the Divine: what is the Divine like, and why? To justify the elevation of purpose to the level of the Divine, for instance, we could show how and why the Divine might exhibit this human attribute and, if the Divine did display it, reveal the divine equivalents to the biochemical and other physical aspects of purpose. Answers will avoid assuming without justification those human qualities for the Divine that require biochemicals.

A related but more central challenge asks how the Divine interacts with the universe. It calls for an alternative to the traditional belief of separation and intervention, a fresh vision for a new or radically revised system of spiritual ideas. The solution this book develops begins where Wright is right: science describes how the Divine works. We work from there.

Why?
if the Divine
wanted us to have
something, we'd need
the biochemicals
for 'em ...

CHAPTER TWO

CONNECTION

Asking Interdisciplinary Questions

As a doctoral student in "Science, Philosophy, and Religion," my aim was to fulfill requirements in the discipline I now call "science and spiritual thought": its history, methods, and points of growth. But the design of the program worked differently from that. It required me to fulfill doctoral requirements in each of the various disciplines from which mine drew: philosophy, physics, mathematics, and religious studies. While the program sounded interdisciplinary, it was instead multidisciplinary. It reflected our culture's tendency to divide knowledge into many disciplines, including spiritual thought and the sciences, and keep them separate. In doing so, our culture houses the study of the Divine away from investigations of the universe.

The reality of the Divine suffers with these disjoint separations. Reality requires the mutual relevance of scientific and spiritual thought. It demands an interdisciplinary approach in which everything connects with everything else, but each still maintains its separate emphasis and purpose. How can connection exist where separateness now rules, and how can it do so while retaining distinctiveness? Let us see.

Keeping Pace with Scientific Theories

If a car leaves Concord, New Hampshire, at seven o'clock to travel to Boston, eighty miles away, how fast must it travel, on average, to arrive at twenty minutes past eight? School lessons abound with such applications of classical physics and mechanics, built on the ideas of Copernicus,

Kepler, and Galileo and later propelled to fame by the physics of Newton and the philosophy of Descartes. According to classical science, to understand something, you need only know the position of its parts at consecutive moments. These are the only uncertain factors—the components' initial velocities and positions. Classical laws explain the rest. They describe the object's movement in relation to all the other objects in the universe and determine its future precisely. Classical physics also holds to the independence of the parts that comprise the universe. They move in a vacuum or a void; at their smallest, they are points on a line with locations but no volume; and they interact mechanically with "pushes and pulls." The Newtonian approach believes in an absolute and universal time independent of space. Time keeps ticking at a constant rate, no matter where in space the clock sits or how fast it moves. The language that describes this time and space, the "order" of classical physics—to use David Bohm's term—developed from Euclidean geometry.

Each period of history develops its own view of the universe to produce its sense of order. The order of Greek science had to do with perfection right out to the circles of the heavens. The clockwork Newtonian order, centered on mechanical movement and expressed by Cartesian coordinates, replaced the Greek order and held sway for several centuries. It still does in most situations. Despite their revolt against many Newtonian ideas, Albert Einstein's relativity theories employ the Newtonian order, as do several assumptions in the usual approach to quantum physics.

Spiritual thought also depends on this classical Newtonian physics. Robert Russell suggests that spiritual thinkers stick with this classical philosophy and outlook because they believe it has received science's stamp of experimental proof. Spiritual thinkers resist working with, let alone building on, more recent theories, fearing those theories may yet turn out to be wrong. Those new theories exist, however, because the old theories ran up against too many problems. Most of quantum theory is far from Cartesian. The highlights of twentieth-century physics and the strange ideas it festoons move away from the classical approach. Now, they tell us, we can't know both the position and velocity of an elementary particle at the same time; one excludes the other. Newtonian physics still holds a place, but within limits, and can no longer function as foundational belief.

Identifying Correlations

We need a new system of spiritual thought, one that, like its scientific counterpart, leans lightly on classical ideas. Mutual relevance steps up as a way to achieve this. Its ideal avoids the classical, disjointed foundations of spiritual and scientific thought. It also raises a challenge for us, one for which the rich store of ideas in physics may contain an answer: perhaps some physical concept can tell us how connection exists where separateness now rules, and exists while retaining the separateness.

John Bell of CERN (Conseil Européen pour la Recherche Nucléaire, the European Council for Nuclear Research) in Geneva restates this usual belief in locality: according to Western philosophy and science, "locality" represents the only rational option: "Consequences . . . don't leap over distances."[1] What happens in one place has nothing to do with what happens at the same moment at some distant place. Physics from Newton through Maxwell to the relativity of Einstein depends on this. So do many of the models that try to relate science and the spiritual, the Divine and the universe.

An essential and new feature in the quantum universe is the opposite of locality: *nonlocality*. It contends that a correlation can exist between what happens, at a single moment of time, to objects some distance from each other. Quantum physics moves away from the classical emphasis on separateness by claiming that an event can have instantaneous consequences for another event outside its immediate area.

This insight may help us establish the mutual relevance of science and spiritual thought. To someone such as Newton, only the Divine can transcend time and space to act in a nonlocal way. Nonlocality, however, does resemble the attraction of gravity, which Newton described so decisively, because it causes objects to affect each other over large distances even though nothing normal and physical connects them. Nonetheless, Newton still would reject nonlocality. Historically, physicists hesitate even to consider it.

Einstein and his Princeton colleagues Boris Podolsky and Nathan Rosen introduced the Einstein-Podolsky-Rosen (EPR) experiment to the world in 1935. Waves from their argument still ripple in our minds, bringing nonlocality to the surface. The EPR trio proposed an imaginary procedure to generate simultaneous events that lie some distance apart and experience no physical interaction. When something happens to one of them, something happens to the other at the same moment. A correlation exists between the events; they appear connected.

A simplified description of the EPR experiment—based on its 1951 treatment by Bohm—starts with a particle that isn't spinning. It breaks in two, and the halves head off at 45 degrees to either side of their parent's path, with equal but opposite spins (the conservation of spin dictates that the total spin of the two halves must equal the zero spin of the parent). When the two lie some distance apart, the experimenter reverses the spin of one of them. What does quantum theory predict will happen to the other half particle? Does it anticipate the classical answer that nothing occurs? No, quantum physics concludes that the spin of the second particle simultaneously reverses with the first's.

Such instantaneous correlations arouse public interest. They also entice us to look here for clues to building the mutual relevance of scientific and spiritual thought.

The EPR experiment as I've described it so far only existed in the imaginations of physicists. Bell brought closer the possibility of performing such a test with a 1964 paper that develops a precise and mathematical distinction between the predictions of a classical theory and those of a nonlocality theory. We need only measure this variable. But researchers found it hard to perfect such an experiment. Starting in 1957 with evidence that hinted at nonlocality, the path to a decisive observation took until 1982 to reach its destination, with the genius and patience of a Parisian team under Alain Aspect. Their results confirm quantum correlations over distances up to 85 feet, perhaps to 110 feet, and disprove theories assuming locality and the classical existence of objects like particles. Further experiments support these results, and researchers plan more.

What tells the half particle its sibling is changing spin? The physics says a correlation exists between the spin-changing events (that is, they are nonlocally related), but what causes the correlation? Despite warnings from people such as Oxford physicist Peter Hodgson against drawing philosophical conclusions from the experiment, speculation runs wild. "As physicists we have learned to live with the EPR experiment, but we have never really come to terms with it," conclude Basil Hiley and F. A. M. Frescura, two colleagues of Bohm at Birkbeck College.[2] Other physicists agree: "Either one must totally abandon the realistic philosophy of most working scientists, or dramatically revise our concept of space-time."[3] The EPR results uphold quantum physics and challenge the usual understanding of space, time, and matter. Many suggestions try to explain the correlations.

• Does a connection exist between the simultaneous events, and does this connection cause the correlation? A correlation differs from a connection. "Connection," a common interpretation of the EPR experiment—and one that shows promise for building the mutual relevance of scientific and spiritual thought—suggests that something links them. "Correlation," what the experiment actually shows, suggests a relationship between them without specifying how it comes about. "Connection" steps toward an explanation of the simultaneous events; "correlation" doesn't.

• T. M. Helliwell and D. A. Konkowski, physicists respectively from Harvey Mudd College in California and the University of Texas, ask about influences that travel faster than light. Could some means send a faster-than-light signal between the two particles?

• Even more extreme, some believe only an *instantaneous* signal connecting the two EPR events could correlate them. Only something traveling at an infinite speed would enable the events to happen at the same time. Many physicists, including Einstein, agree.

• Abner Shimony, a physicist and philosopher at Boston University, suggests a property called "passion" that instantaneously matches the behaviors of two particles far apart, involves speeds equal to or less than light's, and interactions via forces unknown to classical physics. He thinks some form of communication exists that differs from information passing as we know it. "Passion without interaction" fails to satisfy Jean-Pierre Vigier, a physicist from the Institut Henri Poincare in Paris.[4]

• Jack Sarfatti, a San Francisco physicist, asks if a faster-than-light transfer of information without signals immediately connects the two particles.

• Several physicists go further. Does each particle somehow "know" what happens to the other? Hiley finds this unattractive.[5]

According to Einstein's relativity theory, signals don't travel faster than light. Nothing can. Like the horizon, the speed of light always lies beyond us; though our speed may keep increasing, we'll never reach it. A reasonable physical theory, therefore, is local. Quantum theory—and the thought experiment that comes from it—consequently predicts something Einstein considered impossible: signals that travel faster than light.

Nonlocality is a disease, according to University of Copenhagen physicist Richard Mattuck. Jean-Marc Levy-Leblond, a theoretical physicist at the University of Nice, writes about "rather weird nonlocal features" that "plague" many theories with "problems even worse than the ones they try to solve."[6] From this point of view, the EPR experiment must have flaws, and quantum theory must avoid nonlocality to free itself from, in Einstein's famous words, these "spooky actions from a distance."

The majority of physicists, however, want nonlocality and only a degree of the realism that sees existence in a classical way. In this case, quantum physics wins and Einstein, Mattuck, Levy-Leblond and their like-minded colleagues lose; nonlocality is real, despite its weirdness.

We don't know which explanation of nonlocality, if any, is correct. Does that render nonlocal correlations "spooky," something we can't explain? We could hold the EPR correlations as primitive givens, accepted without question, without any explanation possible. According to Martin Dudziak of the Virginia Commonwealth University, conservation laws correlate the behavior of objects over distances and in this way act like nonlocality. In describing the EPR experiment, I applied a conservation law, noting that the conservation of spin compels the total spin of the half particles to equal the spin of their parent. As far as we know, no forces or connections cause this to happen. Nonlocality is as mysterious as the perfectly acceptable and nonspooky conservation laws.

Acknowledging an Underlying Wholeness

The mysteriousness of nonlocality only incites physicists to solve the mystery. How might we explain the nonlocal correlations? We seek a mutual relevance between science and spiritual thought in which they connect intimately with each other yet retain their own identities. The EPR experiment covers a similar situation: a correlation between noninteracting and separate particles. Thus, an exploration of the EPR experiment and its meaning may show us how to create mutual relevance. How might we build it?

An engine makes more sense in a vehicle than permanently mounted and running on a bench. Similarly, the math and the bare EPR facts require a vehicle to give them meaning. Bohm's inquisitiveness carries him deeper than the EPR experiment and quantum nonlocality to develop general ideas that explain the correlations and bestow meaning on the physics.

Between two points on a line, you can find another; between two numbers, you can extract a third. Classical thought, including locality, divides everything into smaller and smaller parts. But quantum physics and its nonlocality say you can't break them up forever. If you keep splitting something, you lose it; if you isolate a piece of reality, you distort its true character. Everything relates to everything else. This quantum "wholeness" is a basic property of the universe.

The experimenter and the subject of the experiment are linked. Starting in the early days of quantum physics, Niels Bohr emphasized this wholeness of a quantum situation. His radically un-Newtonian idea attracts Bohm, who believes that everything connects. For him, the universe is an undivided whole. However, like Vladimir Lenin and Leon Trotsky in the communist revolution in Russia, Bohr and Bohm both rebelled against the classical views only to discover that their versions of the new way led them in different directions. To Bohm's mind, the usual Bohr-inspired quantum physics applies the wholeness unevenly; physicists have yet to take the revolution as seriously as they might. The accepted physics consists of partly related ideas without an underlying and consistent worldview.

Bohm suggests his theory of the "implicate order" as a solution to these problems. Whatever exists lies within the implicate order, he writes, from whatever level it comes—from the quantum world to our macroscopic environment and beyond. The implicate order stands behind or under all of reality.

Bohm explains his theory with the hologram, a device also employed by one of the world's greatest entertainers. Many of us enjoy Disneyland's ghost house with its holographic spook sitting in the cab beside us. Despite the mysterious quality of such images, the Magic Kingdom applies a well-known technique to obtain them:

- Take a beam of laser light, reflect part of it off a mirror, and shine the rest on an object.
- Put a camera where the beam reflected from the mirror meets the light coming off the object.
- The two beams interfere and interact with each other, and the camera captures on a photographic plate the pattern that forms. This plate is the hologram.

- Now light the hologram with another laser, and you produce an image of the object that has three dimensions.

Disney's ghoul enters our cab.

Suppose someone broke the photographic plate from which Disneyland projects the ghostly companion. The theme park operators need only pick up a broken piece of the plate, shine the laser through it, and the whole of the fiend would reappear. Any part of a hologram contains information on the whole object because each portion of the interference pattern from the laser light carries information on the whole object. (Of course, they'd want to pick up the biggest piece, because the smaller the section illuminated, the less detail that emerges.) In comparison, projecting part of a regular photographic image—like one of the negatives returned with your processed film—would produce just a tiny part of the spook, its left ear perhaps.

The hologram helps explain Bohm's theory. Like the information stored in the holographic plate, each region of space and time contains the whole, the total order of the universe, including the past, the present, and the future. Any part of the underlying implicate order carries information on the universe and all its aspects. Bohm refers us to the root of the verb *to implicate:* it means "to enfold" or "fold inward." "Reality as implicate" indicates that everything folds into everything.

We want to apply this approach to nonlocality and the EPR correlations. Everything connects to some extent with everything else, because everything folds into every other thing. Thus the underlying implicate order associates simultaneous but distant events in a way that unifies them; they possess a common foundation that correlates them. No direct causal connection links them. Neither does one event influence another. Rather, the underlying order unites them as it unites everything in space and time. Flick the switch, and lights appear in different parts of the room at the same time. A gigantic conference call simultaneously passes information to each person on the line through an instantaneous connection that would turn AT&T green. Nonlocality expresses the wholeness of the quantum realm, which in turn expresses the wholeness of the underlying order.

Because holism saturates the underlying order, strange things happen to it. To start with, nothing limits it; all limits lie within it. You can't define or measure the underlying order, because if you specify it or describe its content, you divide it. Some things would be in and others out, whereas the underlying order contains all. Neither can you say the underlying order occurs in different forms. Unlike anything else we experience, it contains all its forms.

The mechanistic approach to physics raises the image of the spectacled scientist in a white coat: pens in breast pocket, quiet, collected, and objective. This emphasis on the objectivity of uninvolved and distant physicists can become an authoritarian faith. Finding this outlook inadequate, Bohm opposes it with one that involves persons and emphasizes relationships. He sees a place for both the personal and the uninvolved. The stress on connections appears, therefore, in Bohm's physics, his philosophy, and in the way he thinks physics should proceed. Wholeness dislikes the deification of physics—for instance, its attempt to find a "theory of everything"; physics stands, as does any other form of knowledge, open-ended and incomplete. In principle it fails to describe or comprehend everything.

All theories fall short. Each physical theory straddles a part of reality, and we need all of them. (Bohm thus acknowledges the partial nature of his theories.) Bohm also dislikes a rigid separation between disciplines. The holistic emphasis requires an openness between fields, including spiritual thought and science. It requires that we bridge them. To employ them together and to relate them lead to a fuller understanding of the universe.

The theory of the underlying order emphasizes connection over separation. In the relationship between scientific and spiritual ideas, it may also help us emphasize connection over separation.

Bohm's theory could explain quantum nonlocality, the EPR experiments, and fulfill the mutual relevance of scientific and spiritual thought and still be wrong! The prowess of a theory to explain nonlocality won't clinch its truthfulness, and other offered explanations offer their advantages. Further, no clear experimental support at present gives Bohm's theory of an underlying order, or may ever provide it with an edge over the usual approach to quantum theory. Does this philosophy stand on shaky ground because experiments fail to support or disprove its physics? Because the general theory exists independently from the physics, this is unlikely. We choose this theory for philosophical reasons, because it provides a more coherent account of present knowledge.

If both science and spiritual thought refer to the underlying order, then all aspects of both fields relate to one other. The theory of the underlying order lays the groundwork for mutual relevance. Can we develop this groundwork into a full-blown and flourishing model?

CHAPTER THREE

SEPARATION WITH CONNECTION

Reconciling Wholeness and Division

DAVID BOHM'S MODEL STRETCHES BEYOND PHYSICS to become a general philosophy, a theory of the principles that underlie the universe and human existence. Physicists can apply it to the quantum world, while others can try to use it for other matters. Here, we want to apply it spiritually, to extend what the model teaches to a relationship between the Divine and the universe.

The underlying order of wholeness stands behind or under all of reality. Any part of it carries information on the universe and all its aspects. It boasts the wholeness of all things, total connection. Thus it suggests links where now pervade the distinctions we perceive. We usually experience the parts of the universe as separate and independent (a table, for instance, differs from a chair). How might we instead reconcile the ideas of wholeness and division? Bohm's model answers this and may thus help us develop an image that resolves the split between the Divine and the universe. But does it retain the essentials of the Divine–universe difference? Can it depict the connections where now distinctions rule—without losing the sense of separateness?

In the British Museum, a device slowly stirs a spot of dye into glycerol until the spot disappears. When the machine stirs in the reverse direction, the spot reappears almost to its original state. Bohm calls this "folding" (or "enfolding") the dye into the glycerol, and "unfolding" it out again. The state where glycerol enfolds the dye represents the underlying order of wholeness. The movement or flux of this order involves the unfolding in which certain aspects of it lift into attention, rise into relief. The dye spot reappears in the glycerol. Bohm calls what the unfolding produces the "explicate order," the seemingly stable and independent universe that we experience. It's the order of classical physics.

The spot continuously reappears and disappears into the glycerol as the machine stirs one way and then the other. Likewise, not only does the underlying order of wholeness continuously unfold to become the order of separation that we experience, but also the latter continuously folds back into what underlies it—another part of the movement. If it didn't fold back, endless near-replications of an object from past moments would stand beside each other. Giving reality to something in a continuous way could only happen if the something also incessantly lost its existence.

Challenging the Object-Based View

To chop up a table with an ax scatters it into unordered and separate splinters of wood. The form of a table only alters if you break it up. We usually see the universe as a giant mechanical clock in which each part interacts with the others by pushes and pulls. This primary image of classical physics permeates Western culture. We usually think change arises from the rearrangement of simple parts, the basic building blocks of an object. Grammar also mirrors this object worldview that, according to Bohm, our culture conditions us to accept. The noun, which indicates an object—"I sit at the *table*"—plays a primary role. The verb, which calls attention to action—"I *sit* at the table"—inherits a secondary status. Classically, we prize objects more than activity. Classically, we focus on the spots in the glycerol device.

According to the psychological and neurological research Bohm calls upon, we learn the commonsense idea of unchanging objects in early childhood. We also learn to think of objects as primary. Our culture instills such tactics to help our minds create simpler and more stable forms from the confusion of movements we sense. The forms then build into the objects we see as relatively fixed or slowly moving. Bohm challenges the cultural adaptation or illusion that boils reality down to basic objects, substances, or entities that remain static, rigid, and permanent.

The major point about a hologram, according to Bohm, isn't the photographic plate and its wholeness; it's that movement always takes place. Light waves from the laser continuously interfere with those reflected off the object. This forms an interference pattern, a moving web of light waves whose changing pattern the holographic plate captures at a moment of time in a region of space. Thus Bohm identifies two essential properties of the underlying order. One is wholeness: all parts of the universe, while appearing separate, connect with each other. They emerge from the order without time and space and so in an essential way unite throughout time and space. The other is its constant movement, the stirring of the spots in and out of the glycerol.

Movement precedes stable objects: "Rocks, trees, people, electrons, atoms, planets, galaxies, are . . . the centers or foci of vast processes, extending ultimately over the whole universe."[1] Everything constantly changes, and each hub of change directs us to an overall activity in the universe. With his emphasis on movement, Bohm claims that reality embodies creative activity, but as a flux that goes beyond the motion and transitions of objects. Aspects of the underlying order lift into attention, rise into relief. The underlying order continuously unfolds the order that we can experience, the seemingly stable and independent universe. And once unfolded, this universe of separated things then continuously folds back or enfolds into the underlying wholeness. Activity, change, breaks in regular arrangements are more primitive. We should grant the main role to the action of the verb, Bohm says; we can create the nouns from the verbs.

Each piece of reality constantly forms, reforms, transforms, and ceases to exist, just as the spots in the glycerol appear and then change and disappear as the stirring continues. Every observable thing, Bohm says, "comes into existence, . . . remains relatively stable for some time, and then passes out of existence."[2] Moment by moment, reality precipitates, dissolves, and reprecipitates itself.

Glimpsing the Underlying Order

When you portray a writer at work, you ignore her or his prowess at fishing. When you describe something, you begin with the underlying order. You draw from it a situation broad enough to describe the thing adequately. The context itself plays an active role in unfolding the pertinent aspects of the underlying order; the rest remain unfolded. Further, for a particular situation only one order of experience can emerge at a time. Can a fish swimming in a stream concurrently sizzle in a frying pan? We usually glimpse the underlying order only through its shadow, the order that we experience. Some of it hides from us.

What creates the permanence of a rock? With all this flux and dissolution, why do we sense an object as an object? Because it possesses "continuous form," because it adopts nearly the same configuration in its unfoldings. The cells comprising the human body constantly change, yet we seldom notice this and so we think of it as an enduring object. Anything from a particle on upwards in complexity appears as a semi-independent, near-stable movement of unfoldings.

What happens to the wholeness of the underlying order when it unfolds? The unfolding hides aspects of the wholeness. Thus we experience the universe as incomplete and made up of independent parts. When I tell my physician about a pain I feel, I don't communicate to him what happens in my soul. His interest lies in my physical symptoms: Where do I feel the discomfort? When do I feel it? Does it hurt when I do this or that? The unfolded order is the extrusion of the underlying order of wholeness into a particular form, just as the individuality of a flowering tree's leaves and petals derives from and expresses their common branches, twigs, and flower centers.

Similarly, we often break up objects into approximations adequate for our purposes. As a result the classical approach, including allopathic medicine, wins its place in treating my pain. A revolution for medicine, however, says we should treat a body as more than a collection of machines and recognize that the way my mind handles pain and stress influences the hurt I feel. Separating things into parts whose relationships avoid the whole reaches a limit, because holistic interactions become important at some point. In places, the unfolded retains the wholeness of the underlying order, as we witness with quantum non-locality. Holonomy (the law of the whole) restrains the breaking of a situation into independent parts; pieces arise from the basic whole and in the end relate to it.

The unfolding-enfolding image hosts both connection and distinction without sacrificing either: it maintains distinction because the spots unfold as spots and upholds connection because the spots enfold and become indistinguishable from the glycerol. This makes it a likely candidate to inject life into the mutual relevance of scientific and spiritual thought with its twin emphases on connection and separation.

The waves in this figure join together as one continuous movement.

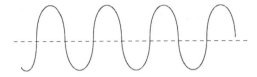

But in the following drawing of the crests of the waves (the portion above the dashed line in the previous rendition), the waves appear separate.

The unfolded order of our experience approximates the underlying order of wholeness by hiding the lower connections beneath an imaginary card it pulls up to the dashed line.

Like an unseen octopus, the underlying order holds all parts of the universe with its tentacles. It founds the wholeness. Yet though essentially indivisible and totally connected, reality usually enters our experience in more or less separate units. The underlying order combines the properties of apparently separate and independent objects by carrying and unfolding to them simultaneous information on the universe and everything in it. The whole imparts to all its members the interactions between them. And it does this without submerging the individuality of each part. Each tentacle of the octopus can be clearly distinguished. The unfolding-enfolding theory thus proclaims both connection (the underlying order) and separation (the unfolded order of our experience).

The underlying order isn't independent of the universe as we can know it. It's both it and it's more than it. It's another way to look at the reality we experience, a way that emphasizes experiences other than separateness and what the eyes of separateness approximate.

To launch this book's system of spiritual ideas with its emphasis on the reality of the Divine, we call on the theme of an underlying whole that produces all things yet, as an approximation, keeps them separate. Similarly, we can separate spiritual experience from the scientific and secular, but only as an approximation. This approach may help resolve but also retain the separation of the Divine from the universe. Each type of experience and knowledge plays its role in the whole, where all have mutual relevance. Bohm's work gives rise to spiritual visions.

Exercising Caution

With the enfolding-unfolding theory, we want to build carefully the mutual relevance of science and spiritual thought. The connections between the parts of the two fields emerge from the underlying order's wholeness from which they all unfold. The separations between the parts emerge from the division of the reality we experience into individual units. The pieces together approximate the original wholeness.

But we ought to bear a caution in mind. Suppose you enjoy Bohm's theory of the unfolding and enfolding of the underlying order and the music of nonlocality, and you wish to apply it to all situations at any level of reality. You might think, for example, that a person in China can connect instantly with someone in the United States. As physicists do with quantum nonlocality, the proponents of such "universal nonlocality" believe it correlates instantaneously and that it defies normal explanations. They maintain that everything correlates with everything else regardless of their separations in time and space and whether they lie at the quantum level (the usual place where physics applies nonlocality). Many proponents invoke this in their spiritual and philosophical uses of nonlocality.

If you know quantum physics, you'll point out that nonlocality acts everywhere at all levels because the theory applies universally. You'd admit, though, that its effects appear in properties of the quantum level. Experiments show nonlocal correlations to happen between the quantum components of objects under certain circumstances, for example, over short distances and for simple systems. It also occurs in complex systems at the quantum level and over longer distances with the temperature near absolute zero (-273.15° centigrade, -459.67° Fahrenheit). Nonlocality only emerges, Bohm and Basil Hiley believe, "in very subtle ways."[3] Nonlocality may exist universally, but experiments in physics have yet to support this extension. Thus the caution: we need to handle the idea of universal nonlocality carefully, because it claims its effects arise at our macro level. The universal variety employs nonlocality as an analogy; physical theory may never establish universal nonlocality as a scientific "fact" of nature.

One theory of the paranormal requires nonlocal action-at-a-distance, in the universal sense, to understand how a psychic in New York can bend spoons in Sydney. A physicist with the U.S. Army, Evan Walker, writes that such a theory of consciousness needs the traits Bohm ascribes to the underlying order of wholeness. The prestigious science journal *Nature* published an article by Bohm and others that endorses a positive approach to the paranormal. Bohm would back Walker's proposal and, it seems, support the idea of universal nonlocality—and yet also recognize the limits of nonlocality.

Bohm writes, "I don't think we should just simply accept something because it's in physics."[4] The 1980 article that carries this statement concerns the parallels between physics and psychology. In it, Bohm also points to problems with a psychological theory that relies heavily on physics. Though the underlying order of wholeness depicts the mind better than the suggestions of older theories, psychologists and—Bohm would add—other scholars should "develop their own basic notions of order to free them from the old ones."[5] People who claim excessive authority from physics for their extensions of Bohm's ideas into other areas are rightly irritated. Robert Crease, a philosopher at the State University of New York at Stony Brook, and science writer Charles Mann note:

> Bell's theorem has acquired near-religious status among certain popular authors who feel it proves the ability of subatomic particles to "think," the basic "wholeness" of the universe, faster-than-light communication, and a host of other mystical foofaraw—what the late great physicist Richard Feynman called the "cargo cults of science." Experiments proving Bell's theorem, wrote Michael Talbot recently in *[Beyond the Quantum]*, constitute "the final proof that reality as we know it doesn't exist at the subatomic level."[6]

To rephrase Crease and Mann mildly, Talbot is wrong. Such abuses of physics appall them. Bohm would approve the caution about overreaching claims from physics, yet he would also accept the idea of universal nonlocality. One ends up uncertain as to what applications of his theories he would support.

John Taylor calls Bohm a modern scientific guru and likens him to the alchemists of the Middle Ages who, to gain immortality, zealously hunted for something that would change a common metal into gold. They wrote in obscure allegories and metaphors to prevent outsiders from understanding their writings and latching onto their secrets. Taylor suggests that Bohm similarly employs obscurity enhanced with scientific ideas and their extensions. Remote and authoritative to the public, Bohm's words inspire as much awe and mystical reverence as the alchemists' words did to the people of old. The comparison hardens when modern scientific gurus answer the big questions—"Why am I here? What is the purpose of life?"—with their approaches to science. Bohm extends his physical theories past physics to an all-explaining system of unfolding and enfolding orders. This leads many people to associate Bohm's name with fringe topics and movements. Whatever cautions Bohm proposes, some people discern only the excesses shining through. They win a point.

Taylor insists we hold gurus scientifically accountable for the ideas from which they start and adds that Bohm's beginning point, the unfolding-enfolding theory, glosses over serious flaws. He also requires that we scrutinize where the gurus send these ideas: to the "distant waters of the mind and soul, of ethics, morality."[7] Hiley questions Taylor's analysis of Bohm's physics, but the alchemist point still rings true. Bohm does apply his theory to explain a range of experience that overflows physics. He should deploy the physics correctly and declare that his wider theories leave the safety of physics' support.

One of my students, an ex-Roman Catholic priest, encounters significance in the nearly forgotten symbolism of his church's architecture and esoteric tradition ("the inner, hidden, deeper dimension of reality").[8] Many of my students reject Western religion because of what the institutional church stands for or did to them. Some abandon it while others cautiously seek meaning through it. To search with honesty requires the wisdom of our experience and an awareness of limits. We need to be wary, but persist in our struggle for sense.

The proviso I have just written also applies to the proposal this book develops. Perhaps universal nonlocality embodies or acts for the Divine. Perhaps the Divine is the underlying order, whose unfolded wholeness expresses divinity. These suggestions extend physics; to explore them requires that we remain aware of the caution and refuse to say physics proves them.

Part Two
Searching the Divine

—

CHAPTER FOUR

THE SUBUNIVERSE DIVINE

Finding Language to Describe the Origins of the Universe

COSMOLOGISTS USUALLY TRACE THE ORIGIN OF THE UNIVERSE to the big bang, the gigantic explosion and fireball that began the universe between twelve and twenty billion years ago. But what caused the big bang?

Physicist Edward Tryon of Hunter College in New York suggests the big bang started from events at the quantum level of something called "the vacuum." This vacuum contains nothing, not even space or time. So what can we say about it? Very little. We do know it possesses infinite energy (something physics usually ignores) and that microscopic particles materialize in it and then instantly annihilate each other. One of these fleeting fluctuations flashed up the big bang.

Even if Tryon's idea holds up and the vacuum produced the big bang, darkness still surrounds the origin of the vacuum and the laws by which the quantum universe operates. Big bang cosmology says nothing about their origin. Several physicists imagine, however, that something more basic than the big bang, its products, physical laws, or the vacuum does or did exist. From it, the vacuum and the laws emerged, and these in turn produced the big bang.

Before throwing ourselves into the search for the "something," we need to confront our language. Discussion about the beginning of the universe stands on shaky ground. We want to talk about actions before the big bang and the provision of existence to the universe, but the meaning of such words as *origin, beginning, created, acts, happened,* and *before* arises from the way they apply to the ideas of time and space, in particular the notion of time passing. *Before,* for instance, means "stands in front of" spatially or in time. Since time and space only started with the big bang, the phrase "before the big bang" is senseless. Language breaks down when talking about a supposed something that happened outside of time and space. What does it mean, then, to say that something produced the big bang?

Perhaps the phrase "the origin of the big bang" is meaningless. Should we give up in despair and cease all talk of objects or beings or events outside space and time? No. Though we fumble in our language, we nonetheless should humbly assume we can understand this talk, that it makes sense. But we should proceed with caution. We should tentatively apply words to a situation without space and time only after we study them and how they fit: What do the words mean in the new setting? What meanings fail to carry over to the new situation? What dangers lurk in the new usage? Because of this problem with words, David Bohm studied his language of the underlying order and changed the language as a result. Perhaps proof that we do understand the words and use them correctly will lie in how well our theories hold together. We also can look at the value and soundness of any conclusions that emerge. We can test the language, subject it to experience and reason. The system of spiritual ideas that this book unravels is such an experiment and probe. It accepts the invitation.

Starting with Something

Now we are ready to ask what is that *something* that produced the big bang.

The Restaurant at the Beginning of the Universe doesn't serve a "free lunch"—or breakfast, for that matter. However many times we ask the question of the origin of the universe and however many layers we push it back through, we always end up with something unexplained. Theories leave some matters unjustified; each starts with something. Willem Drees develops this point in his book *Beyond the Big Bang: Quantum Cosmologies and God*. The big bang theory assumes four things about the universe, he tells us: the laws of physics, the three dimensions of space and one of time, the conditions of the universe at its initial moment, and the existence of the universe. The nothing from which the universe arose is, in Drees's words, "not an absolute nothing."[1]

Stephen Hawking employs the anthropic principle to explain the pre–big bang something. At the base of a universe lies a set of laws. Just as changing the combination of ingredients for a cake leads to different tastes and textures, so different sets of laws produce different universes.

The anthropic principle shows that the evolution of creatures able to probe natural laws requires a special environment in the universe and, the Cambridge physicist adds, only one theory of cosmic beginnings or one set of laws leads to a universe with those special settings. Since the inquisitive creatures (namely we) do exist, cosmologists must choose that theory. If we want our cake to taste sweet, we must include in it a form of sugar. Thus this universe could only possess the laws it has. Hawking also concludes—adding a further difference between his theory and the others Drees looks at—that the universe could only start with the initial conditions it did. Physicist David Schramm of the University of Chicago roots for Hawking: "You just have to say: in the beginning before the big bang, there was mathematical consistency. Everything else follows, including us."[2]

Notice that Hawking's argument assumes the preexistence of a form of logic, what Schramm calls "consistency." Hawking's cosmology tweaks the imagination. It also helps solve a problem.

Our usual idea of time and space assumes a shape or structure (a "geometry") for the universe, writes Bohm, that builds on the mathematical idea of a continuum or a line. Remember drawing a line in geometry? You had to keep your pencil on the paper or else you broke the line into several segments. A line flows continuously. It comprises an infinite number of points lying side by side. And remember marking a dot to indicate a point? Though your pencil spot occupied space, it only approximated the real point, because the real point took up zero room. Thus, points can lie very close together and their pencil marks can coincide, but the points remain distinct. The space-time continuum declares that events fall at different points and even those very close together can be separated. On the other hand, quantum theory demands that the opposite pervades the universe: no matter where they lie on a line, no matter how distinct they appear on the sheet of paper, points connect with each other.

I've seen watches advertised that not only tell the hour, minute, second, date, day, and year for here and every conceivable place on the globe, but they also act as a stopwatch, an alarm clock (with several settings per day), and a radio and automatically adjust themselves when daylight savings time starts and finishes. But faced with all those options, I doubt I would ever figure out and remember what all the buttons do and how to operate them; the excess of features would interfere with my basic needs for a watch. The space-time continuum also possesses too many features, and these interfere with attempts to understand what happens from the point of view of quantum physics.

To shop for a pen, you must decide among different colors, widths of line, gold or plastic, short life or long life, three colors in one or a single color, a box of twelve or a single unit. Or you could purchase the basic, no-frills store brand for less than the others. If none of these suit you, you could always grab a piece of charcoal from the remnants of last night's fire. Princetonian John Wheeler introduces the "pregeometry" to replace the continuum with its surplus of features. The pregeometry resembles charcoal in its bare-bones quality; the continuum compares to a name-brand pen with a fluorescent barrel and medium point. A preexisting form of logic, the pregeometry creates a more basic structure or geometry than the continuum, without its problems. Both relativity and quantum theory assume the existence of the pregeometry.

Rather than using Wheeler's penetrating yet speculative word *pregeometry,* I opt for the more neutral and simpler term: *subuniverse.* The pregeometry is a subuniverse. Physicists may disagree with Wheeler over what the big bang requires, whether it calls for a geometry or for, say, a primitive form of logic. They may agree, however, that it requires something and that the something embraces at least the subuniverse. Wheeler travels a different path to the conclusion—the preexistence of a basic form of logic—I draw from Hawking.

Besides a prelogic, something else preexists, as Drees points out:

> Even if theories are perfect and complete, they do not answer the question of why there is anything which behaves according to those theories. The mystery of existence is unassailable. It remains possible, therefore, to understand the Universe as a gift, as grace.[3]

Something bestows existence on the universe, raising it from a conjecture that may or may not happen to a reality. Hawking similarly notes,

> Even if there is only one possible unified theory, it is just a set of rules and equations. What is it that breathes fire into the equations and makes a universe for them to describe? The usual approach of science of constructing a mathematical model cannot answer the questions of why there should be a universe for the model to describe.[4]

The mystery—the gift of which Drees writes and the fire to which Hawking refers—plus a preexisting logic from which the laws of the universe develop lie in the subuniverse. They provide the universe, Wheeler writes, "with a way to come into being."[5] The subuniverse gives existence to the universe, and its actions are the laws of nature.

Exploring the Subuniverse

A child, once born, can live without its mother. The subuniverse conferred existence on the universe, but then what happened to the subuniverse? We usually think the universe grants itself existence through time and imparts reality to its laws, that it, in other words, assumes the function of the subuniverse. Did the subuniverse shut down with the big bang? To survive, a person—a baby, especially—needs other people. Perhaps the universe similarly requires its subuniverse. After all, time and space don't apply to the subuniverse, because it exists outside them and creates them. Public resolve and money set up the local fire brigade, and it will exist and operate for the foreseeable future because the original act includes daily financial backing. The subuniverse's initial provision of existence is the same as supplying it throughout time. One act covers all time because the act continues on forever. The subuniverse produced the universe and its parts. It continues to give them existence moment by moment, and its continuing acts of empowerment we describe as the "rule of natural laws." The subuniverse pumps as the heart of the universe.

Besides a statement like this, what words describe the subuniverse? In several articles, Bohm and Basil Hiley show how the underlying order of wholeness satisfies Wheeler's search for a pregeometry. The order produces existence for the universe and everything in it and operates by logic; as the universe's mother, the underlying subuniverse continuously gives it birth and rationality throughout time. It may provide it with more as well. You might recall these Ash Wednesday words: "Remember thou art but dust, and to dust thou shalt return." The underlying subuniverse forms the dust to which the universe we experience returns for ceaseless resurrection. The unfolding-enfolding theory provides a rich language for exploring how the subuniverse created the big bang and continues to endow the universe with existence.

Relating the Universe and the Divine

As the universe derives from the Divine, so many spiritual traditions would say, it reflects directly and intimately basic characteristics of the Divine. We can apply this to the laws or reasoning of the universe and the Divine. As the features of the pot reflect the mind of the potter, the basic laws of the universe arise from the Divine and reflect the Divine's own reasoning. Spiritual tradition associates creation with the rational mind of the Divine. Thomas Torrance, a Scot, emphasizes something else as well. The universe arises from not only the Divine's reasoning but also, Torrance points out, from the Divine's creative power. The existence of the pot, not just its features, reflects the ability of the potter to construct from raw materials. In Hebrew, the name of the creator is YHWH (Yahweh, or Jehovah) or, translated into English, the one who "brings into existence whatever exists." These two properties of the Divine, logic and fruitfulness, are properties of the subuniverse. The subuniverse mirrors the essential properties of the Divine.

Like many others, you may think the Divine initially created the universe from nothing, say through the big bang. Spelling out this doctrine, however, leads to debate, even hostility, as in the creationism fracas. What does "from nothing" mean? Did the Divine form it from primordial dust, the void of Genesis 1, back in 4004 B.C.E. (the date of creation the Irish prelate, James Usher, worked out in 1650)? Or is our universe a bounce back from the big crunch at the conclusion of a prior universe fifteen billion years ago? Did it slither out of a wormhole connecting it to another universe? Beneath these disputes, believers do agree on something: the universe and everything in it depend for their original existence on the Divine. Everything is, in a word from tradition, "contingent": the Divine decided to produce them and did. The subuniverse approach says much the same: the subuniverse gave existence to the background vacuum and the laws of quantum physics, and thus to the universe.

"The Divine creates out of nothing at the beginning." "The Divine continuously accords existence to the created universe, moment by moment." In many belief systems, the creator role of the Divine divides in two: the creator and the sustainer. "To sustain," defines Webster's, is "to keep in existence."

Nothing in nature arises out of nothing. Everything in it emanates from something else, the product of strings of generations, each of which unfolds from the underlying subuniverse. Robert Russell suggests that this idea of Bohm resembles the spiritual belief that everything depends for its existence on the Divine's sustaining power. Just as anything in the universe of our experience exists because of the continuous unfolding of the underlying subuniverse, so, for believers, the existence of anything relies on the continuous and creative activity of the Divine as its sustainer. The Divine causes the universe to exist and to perpetuate. The Divine continually produces each item, relationship, and feeling; the Divine carries out everything, produces everything (subjective, objective, or from any other category), gives it existence moment by moment, and is responsible for all changes or nonchanges in each object and system, moment by moment. The Divine achieves this by unfolding the potential that is of the Divine. It pumps as the heart of the universe. Thus the suggestion that the Divine is the subuniverse yields the traditional doctrines of original creation and of continuous creating and sustaining. It wraps flesh around the word *sustainer.*

But, you may think, events happen because of natural laws. The universe is a self-determining, self-empowering, self-existing, and self-perpetuating entity; it grants itself existence through time and imparts reality to its laws. It carries out everything itself. You may concede that a subuniverse performs these functions, but still you would think of it as natural.

So what causes the universe to exist and behave as it does: the universe itself, or the divinity as the subuniverse, unfolding existence through every moment of time and point in space? Both. The subuniverse Divine is that aspect of the universe that gives it existence and causes it to behave consistently. The force of natural laws lies in their attachment to the subuniverse Divine. They don't refer to Platonic-like powers that exist at a different level than the universe. We need to evict the separation between the Divine and the universe and thus quash the dual agent answer to the question of what brings everything about. Even the usual understanding of the sustaining function of the Divine enshrines this separation, this philosophy of independent objects, because it thinks the Divine holds things in existence while remaining apart from them. We incline toward the image of ourselves acting on objects outside of ourselves. My fingers touch the keys of my computer, for instance. But what if we think of my fingers operating on their own? They then become an aspect of the keys and my thoughts transfer directly from my mind to the screen. Similarly, the Divine acts in a lawlike manner to unfold from the divine self the regularities in how things affect each other.

The subuniverse divinity resembles a basic hamburger: just the meat and plain white bun, no ketchup, pickles, sesame seeds, mayonnaise, lettuce, tomatoes, relish, onions, pineapple, or cheese. At this point, I abstain from adding meanings to the basic functions for the Divine (that the subuniverse operates according to a form of logic and that it dispenses reality to the universe and effectiveness to its laws). I abstain from adding meanings such as the Christian belief that the creator becomes a particular human being, Jesus Christ. I start from scratch.

But does this generate a redundant idea for the Divine? If the word applies to everything, then it's empty, a tautology like the statement "all water is wet." It says little to claim the Divine causes everything. Does it make spiritual ideas, as the Dane, Viggo Mortensen, phrases it: "nothing but words—words that we could just as well do without"?[6] Though the idea of the Divine and the idea of the subuniverse fit harmoniously, would that of the universe suffice?

A hundred years ago, the idea of ATM cards made no sense. Slide a plastic (what's that?) card with a magnetic (magnetic?) strip into a machine and money appears? It knows the balance in your account? Circumstances change and, along with them, our experience and our language. The lack of language for something says naught about its existence.

This is also true for the image of what underlies the universe of our experience. To offer meaning for our lives, the idea of the Divine requires more than the creator and sustainer functions. A basic hamburger may knock the edge off your appetite, but it will fail to zing your taste buds. It's difficult at the moment to say exactly what will emerge as we expand the language. If we slowly build an appropriate language, matching the growing experience of the reality with our words and ideas, the subuniverse picture of the Divine may turn out as real and as useful as ATMs and plastic cards. Far from superfluous, as Mortensen might claim, the idea may inform and include more functions (inspiring worship and grounding values, for example) than could scientific theories. Such spice for the creator assumption will emerge as we further explore the Divine.

CHAPTER FIVE

IT JUST IS

Facing the Challenge of No Beginning

THE UPSTART IDEA THAT THE DIVINE IS THE SUBUNIVERSE faces serious challenges. To understand them, let's consider again Stephen Hawking's cosmology. As we noted in chapter 4, his cosmology seems to assume the existence of something that produced the big bang. That something created the big bang and moment by moment continues to create the universe, the world of our experience. It operates logically and dispenses reality to the universe and effectiveness to its laws.

We call this something the "subuniverse." Tradition ascribes these properties to the crux of the Divine. Thus we credit the parenthood of the universe to the Divine: the subuniverse is the Divine.

But let's imagine the universe as a cone, with its peak at the top. Beginning at the bottom, we trace the universe back through time toward the big bang, toward the top of the cone. In Hawking's cosmology, rather than reaching a point, the cone rounds off like the top of a sphere. A sphere has no top point. Similarly, according to Hawking, no point can represent the big bang and the beginning of time. The universe has no beginning.

No start to the big bang? No original creation? Here the subuniverse understanding of the Divine confronts the might of Hawking's speculation. Yet since the subuniverse exists out of time, our theory avoids the pressure for a beginning point of the universe and conforms to Hawking's theory. At the same time, the subuniverse does not rule out a beginning point; it could create one. (Think about the effect of laws of one country on citizens of another: while I reside in the United States, for instance, Spain does not require me to obey its laws.) Thus, the subuniverse idea fits with Hawking's cosmology, and it fits with a theory that represents the big bang with a point. His claim leaves our idea intact. So much for the first challenge.

Facing the Challenge to the Existence of the Divine

Spiritual people may believe the Divine chose the characteristics of the universe from an array of alternatives whose options we can only start to imagine, an array that puts the Mall of America to shame. Usually, the spiritually minded believe the Divine acted freely to create the universe, choosing the initial conditions for the universe and its laws. Hawking's "no beginning point" leads to a more serious challenge than whether that fits with the subuniverse idea. It forecloses the mall. That his universe avoids a start-up moment means the Divine had no choice in its initial settings. On top of this, the anthropic principle shows that only one set of laws can lead to a universe with an environment suited to our evolution. The laws the universe follows probably weren't up for grabs either. The universe is self-contained. It just is. Nothing can create or destroy it, and nothing outside it can affect it. Hawking puts the Divine out of a job.

He ups the tension between spiritual belief and atheism. You sometimes hear that science can't prove the existence of the Divine, but here Hawking compels cosmology to undermine the Divine's existence. Does Hawking do away with any attempt to introduce the Divine into the picture? I claim the opposite is true of his physics: it actually builds on the assumption that the Divine exists.

To look at it another way, we might ask what divinity Hawking's belief renders redundant. Ted Peters, of Pacific Lutheran Seminary in Berkeley, thinks Hawking's interest lies in deism, a form of spiritual thought in which the Divine brought the universe into existence and then went on vacation. The abandoned child runs on its inborn laws. If cosmology supports Hawking's theory that the creation of the universe needs only itself, Peters agrees that science ousts a *deist* creator. Theists, on the other hand, believe the Divine performed something at the beginning and remains active through time. British physicist Chris Isham compares the actions of the theist's Divine with quantum processes (especially the collapse of the wave function) that continuously bring something into existence out of nothing; from mere possibilities a reality emerges. Similarly, says Isham, the Divine acts now. The creative work of the Divine pushes on beyond a single event that happened way back in the past. It keeps going. Peters points out that theism replaced deism in most spiritual circles some time ago and that, for some reason, Hawking passes over the usual understanding of the Divine for the superseded, deistic model.

You might think Hawking downsizes heaven and earth. His cosmology has, in Pierre Simon de Laplace's sentiments, "no need of the Divine hypothesis." Yet we could replace *the Divine* in the above thoughts from Isham with the *subuniverse,* without losing his meaning. The issue is more complex than Peters recognizes because he overlooks several of Hawking's assumptions about the universe, especially what I call "the subuniverse and its relationship to the Divine." Yes, Hawking's cosmology does shed deism, but, with the idea of the subuniverse, it attires itself with theism. Despite the atheistic bantering, Hawking's cosmology depends on the existence of a form of the Divine.

Facing the Challenge of Divine Origins

More obstacles rise into view. I say the universe unfolds from the subuniverse. The child in us asks for the origin of the subuniverse. I suggest the Divine as the answer. The wise still ask what produced the Divine. I fumble for an answer. As another objection to the notion of a divine subuniverse, you might say that some theory could explain the origin of the subuniverse and so declare a divinity superfluous.

To well-educated spiritual thinkers, asking what produced the Divine is to ask a question as improper and unnecessary as, What does a square circle look like? The Divine has no origin, they say. Perhaps this means they don't know the answer or you shouldn't ask. In response to the big question about origins, spiritual wisdom lapses into silence.

Rather than a spiritual concern, could a science such as physics answer it? Only one physical theory explains the big bang origin of the universe, says Hawking, but he wonders if it holds the power to cause existence. We might ask if any theory gains this ability without begging the question. What endows it with the might to enact itself and produce everything? It would—by definition—refer to the Divine if it achieved this by itself. This route leads nowhere.

Another might head us where we want. Hawking's mall closing relies on the anthropic principle. Though he fails to expel the idea of the Divine, the principle may still carry some leverage and help explain the origin of the subuniverse.

Why does light zip along at 670,000,000 miles per hour and not at 55? Why do the constants of the universe, such as the speed of light, possess the particular values they do? Another previously noted observation fuels such questions: the constants must possess these values for intelligent life to exist in the universe. A slight variation in any of them would render the evolution of beings such as us impossible. The universe would live too short at too hot a temperature or expand too slowly with gravity too feeble for planets to form.

Two forms of the anthropic principle start to answer this physical problem. The "weak" version simply and blandly asserts that the constants carry the values they do because we exist. Our universe must suit the evolution of intelligent beings because we evolved here. The "strong" anthropic principle contends that myriad universes occur and that at least one of them carries the values of the constants suited to the evolution of intelligent beings. We, of course, live in that universe. Rather than restating the fact that we evolved in this universe, this version of the principle attempts to explain why intelligent life could develop here. It calls on the idea of many universes to help us understand why the constants of our universe carry their particular values.

Strict science would shun the anthropic principle as invalid and not established. As a way to solve the problem about the constants, it feels temporary and makeshift; it lacks profound physics. But at least it starts an answer.

Might a form of the anthropic principle also help us understand the origin of the subuniverse? The principle may offer the best—perhaps only—approach of physics to this origin question. To see where this leads, let the (pre-)logic of the subuniverse represent it and the laws under which the resulting universe operates. Logic could, conceivably, vary from subuniverse to subuniverse, if more than one does exist. (The other property of the subuniverse—the bringing of the universe and its laws into being— may remain the same, because it either operates or it doesn't; if the latter, the universe doesn't exist and this chapter plus you and I self-destruct.) But another matter invites clarification. Earlier in this chapter, we noted Hawking's use of the anthropic principle to whittle down the number of theories that explain the universe. Consistency and the anthropic principle mean, he thinks, that we need only one theory—and only one can do it anyway. I apply the principle in another way. Different theories all operating under the same logic concern him; different logics concern us.

Does the fact of human existence suggest that the logic of the subuniverse (and the resultant laws of our universe) must be as it is? Does our presence constrain the subuniverse?

Let us take an analogy. In Aztec religion, numerous gods ruled over daily life, including Uitzilopochtli (the sun god) and Coyolxauhqui (the moon goddess). Uitzilopochtli murdered his sister, Coyolxauhqui, and so began a blood-demanding drama of life. As a result, Aztec warriors felt it an ultimate honor to die as ritual sacrifices, ascending the steps of a large pyramid where priests would stretch them across a stone and rip out their hearts.

Aztec beliefs elicited activities we consider illogical, unintelligent, and inhumane. Our culture claims to esteem human life and outlaws sacrificing it on altars. It avows a belief system in which gods behave differently than those of ancient Americans. We might claim that our beliefs show logic, intelligence, and humaneness. Now, suppose that "intelligence" requires our logic A (we define intelligence in terms of logic A, after all) and that there exists a subuniverse with a logic B, different from our logic A. The subuniverse with logic B produces a universe with logic B, and whatever this universe generates operates also with logic B. We could say the mythology of the Aztecs acts like logic B to create a world of meaning foreign to what makes sense to us in our world. Because we with logic A exist in this universe, its subuniverse must follow from A. The weak anthropic principle for the subuniverse states that the subuniverse bears its particular logic because humans live here to observe its product universe. If a different logic possessed our subuniverse, intelligent life wouldn't exist here. By this reasoning, this weak anthropic principle conveys little.

The strong version envisions a host of subuniverses and universes whose logics could be anything. At least one of them has logic A, and, in it, we who operate under A evolved. While this at first makes sense, consider a subtlety: Can we think what a subuniverse with a different logic might produce in a universe? It is hard enough to understand someone speaking another language within the same logic. Mathematicians can construct different logics, but other reasonings lie beyond imagination and we have no idea what such different logics produce. Further, it requires human reasoning to assert that "a subuniverse with a different logic produces a universe." To think this way about other universes calls for a consistency between them and our universe, a type of universal sense. This cuts across the assumption that different subuniverses exist with different logics. If this talk ties you up in knots, join the club; it ties itself up.

An anthropic approach stalls with subuniverses, and current science may fail to explain the origin of the basic laws of nature and hence of the subuniverse. The Austrian American logician, Kurt Gödel, devised a famous theorem that he published in 1931: some of the propositions that found a system of math will remain unproven. In any logical system that employs symbols, we can (or at least Gödel could) construct an axiom that we can neither prove nor disprove within the system. To show the self-consistency of the system requires methods of proof from outside it. This suggests that a logical system such as physics will never explain itself fully. So we should not expect science to account for all its assumptions.

Neither a scientific nor a spiritual approach will understand the origin of the subuniverse. It just is. But we can use Hawking's ideas as training wheels to create a more capable theory.

CHAPTER SIX

THE DIVINE ACTS?

Expecting a Law-Abiding Divinity

I fiRST ENTERED THE UNITED STATES to attend Princeton Theological Seminary. After traveling all day and night, my wife and I arrived at JFK International Airport. We were exhausted, overwhelmed, and confused. No one from the seminary was there to meet us. Maybe we should call them. But how do you call collect? What's a dime? A woman behind us in the line for the telephone asked where we were headed. She and her husband lived in Princeton, and she offered us a ride. I look at this rescue as an action of the Divine in my life.

How did the Divine act in this specific event and send these people to pick us up? Perhaps their phone rang and the Divine told them to meet us. Or perhaps the divine planner had influenced events from way back so their plane landed just after ours and they had to use the pay phone to find out the condition of their daughter who just happened to contract tuber-culosis that morning. The usual approach to the question of miracles is to say they occur by simple and direct intervention into the natural course of events: a large finger points down from the clouds and a deep voice booms.

If a plane leaves London with 10,000 gallons of fuel and needs the energy of 5,000 gallons to fly to JFK, it won't land there with the 10,000 gallons still in its tanks. Suppose we isolate a situation in an imaginary box that energy neither enters nor leaves, or the energy entering equals the energy leaving. Physics calls such boxes "closed systems." A conservation law of physics says the energy in such a system stays the same over time; the plane must expend fuel to fly from London to New York.

Suppose the aircraft in which our rescuers flew to JFK arrived late because the Divine caused the wind to blow against the plane harder than

expected. Blowing harder requires extra energy. Where did this energy for
the plane-hampering wind arise from? Hardly from within the box con-
taining the plane and the air around it. It magically appeared from
nowhere. The intervention approach to miracles presents an energy problem like the plane that flies on nothing. Changes the Divine creates by
intervening in a system require energy that wouldn't occur naturally, and
so they break the conservation laws of science. We want to maneuver
around this difficulty.

We expect the Divine not to mess with the universe in a magical or
capricious way. Humans didn't break into existence with the Divine meddling in the genes of an ape. Neither would the Divine add energy at some
places and overlook others. A rescue here, a healing there; a miracle for me,
a miracle for you; but nothing for a baby run over by a crazy driver. Everything must make sense, somehow. How could science, let alone human
life, exist in an erratic universe? We expect the Divine to interact consistently with the universe and to respect its integrity.

Acting through Chance

Our agenda now lists two types of questions. First, how might we relate the
actions of the infinite to the finite world? Where could they occur? In particular, how might the Divine act in individual and specific ways? Second,
how might the Divine achieve this without stirring up an energy problem
and unleashing an erratic universe? We want to understand the unique
events of modern Western experience that evade scientific explanations of
the regularities of nature and human behavior and yet are real to us.

We might envisage the Divine's special actions as relating closely to
the Divine's regular and dependable interactions with the universe, such as
creating and sustaining it by unfolding it. Of course, the Divine can only
employ unfolding to interact in specific ways if those ways conform to the
laws of nature. Our model of the Divine might show how the Divine acts
in special ways as well as in a regular scientific, lawlike manner.

Like a drill sergeant whose screaming power of abuse forces everyone
into the step he calls, some think the Divine prescribes and compels everything to happen as the divine will dictates. Physicist C. W. Rietdijk
believes the Divine could only create a universe in which everything is
determined. Indeterminism conflicts with the Divine's nature. "Our only

hope of survival in the deepest sense of the word," Rietdijk writes, "the only hope of the truly religious person, has to be set on determinism," in particular on a deterministic quantum physics.[1] In the context of determinism, he argues, we will understand the Divine's special actions.

But many thinkers advocate indeterminism—especially that of the quantum world—as the way to approach divine action. A fully determined nature, according to physicist Frederik Belinfante, prevents the Divine from doing what the Divine might want to do; the way nature operates restricts the options available to the Divine. Physics reveals that not everything is fixed by something else. Uncertainty holds sway at the quantum level. If you know the position of an electron, you can't also know its momentum. So Belinfante tosses out the idea of a fully determined universe and expects divine actions to occur where uncertainty dominates. He continues by explaining how the Divine might use the lack of precision to add an action or two or three into our universe's tangled mesh of events. Another quirk of quantum theory states that before you observe a quantum system, you can only supply the chance, the odds, that the system will possess these or those properties. You can't tell for sure what you'll see. You can't be certain of the results of an unrigged horse race before the gate opens. But the Divine, Belinfante suggests, knows which horse will win, what state the quantum system will become when it changes from a possibility to a reality, when someone observes it. The Divine knows this because the Divine decides what condition the system will adopt when you see it. Other scholars suggest similarly: divine influence could produce one outcome over other possibilities. The Divine continuously decides what happens in the universe, matters you and I can't predict. Thus quantum physics allows for belief in an active divinity.

Interacting with Various Levels?

The Divine, it seems, can act through at least the quantum level of chance. Does the Divine also interact with other levels via those of chance? Do the effects filter into them?

In my office stands a ceiling-high case of shelves, each about three inches high, in which I keep my students' in-process dissertations and learning agreements. Robert Russell recounts that the unfolded universe contains an infinity of levels, and my imagination compares them with these shelves. In some of the levels, determinism predominates and, in others, the uncertainty of quantum physics and roulette wheels holds sway. On my lowest shelf, I file procedural handouts for those about to embark on learning agreements or dissertations. My bottom shelf affects those above it because of this. Russell reminds us that spiritual thinkers frequently place actions of the Divine into a supposed bottom tier of the universe's levels from which effects seep up into our experience. But in reality, no bottom story exists in the universe of our experience, because the chains of levels spread out forever; neither uncertainty nor determinism speaks the first word. Without a lowest shelf, the universe forces the Divine to function through regular levels.

My students' learning agreements greatly influence their dissertations, because the former lay out the plans for the latter. From where they first operate, do the actions of the Divine find their way into and affect other levels? No matter what type of level the Divine acts through—chance based, determined, or anything else—this question persists. Suppose divine actions on one level fail to influence any others because an impervious barrier separates scientific and spiritual theories. Then, because science claims to provide full understandings, the Divine becomes irrelevant to explanations. Science and the Divine sit incommunicado on different shelves—something Russell and I resist. Can we instead show that divine actions do influence other levels? Can we show that a permeable junction lies between spiritual and scientific theories? Ian Barbour points out that the Divine can't reach our experience through quantum uncertainty; no known connection exists between the uncertainty and, for instance, our freewill decisions. We don't know what a relationship between these two levels (or between most others) must entail for one to affect the other. Thus, the levels idea still fails to explain how the Divine acts in or on the universe. The idea of levels malfunctions unless it clearly understands the to-and-fro rapport between the levels.

To claim the Divine operates in certain levels moves the obstacle the Divine needs to cross in order to act; it used to lie on the border between heaven and earth, and now it rests between the levels of the universe. As the Divine "totally beyond" the universe erects an insurmountable barrier between the level of the Divine and that of the universe, so chains of levels can plant nearly impenetrable, maybe impervious, obstructions. This lack of connection between the levels bespeaks a lack of wholeness. Thus we bid farewell to the model of separated levels, and turn to search for one of connections.

Searching for Connections

In the West, we say that tissues determine what happens to organs, and organs determine what happens to the whole organism. The parts of something, we feel sure, determine how it behaves; lower levels rule what upper ones do. Now ask if the organism can influence its heart and lungs, or the lungs their tissues. In other words, turn the upward scheme around. Do higher levels act downward on lower levels? Does a whole modify its parts?

A colleague faced the surgical removal of a cancerous tumor in his neck. Besides conventional medicine, he sought the help of an Eastern healer who taught him how to meditate and so control the flow of blood in the area of the tumor. My friend practiced and practiced until the technique became second nature to him. He required none of the six pints of blood the surgeons had been waiting to transfuse. Stories abound with downward explanations of "mind over matter."

The cerebellum, the cerebral cortex, the hemispheres, the thalamus, the lobes, the neurons, the synapses, and so on all constitute the brain that, as a whole entity, exceeds what its components can achieve. Nobel Prize-winning neuroscientist Roger Sperry develops an image of the downward action from the mind to the body. He describes this as the "brain-as-a-whole," the total state of the brain that surpasses all the components that comprise it. I call it the "Brain," with a capital *B*. When I think to write, "The cow jumped over the moon," a conscious process at the level of my Brain runs. It causes my lower-level neurons to fire off frenetically with signals that travel to my arms, hands, and fingers to type the words. The Brain lies at a level that can control or influence what happens at the lower neuron tier of the brain. Mental events—thoughts, feelings, decisions—aren't events in a mind separate from the physical universe that mysteriously interact with the brain, like a guardian angel. They apply to the Brain, says Sperry.[2] They describe total states of that physical organ.

Sperry's Brain model helps us think about the universe. The Brain forms a whole, the brain-as-a-whole, that exceeds everything in the brain. It refers to the total state of the brain. By analogy, imagine *the Universe,* meaning the "universe-as-a-whole," or the total state of the universe. The

Universe, considered as an entity, forms a whole that exceeds everything in it.

Acting Downwardly

When they chart the course for an interplanetary spacecraft, NASA scientists allow for the push of the sun's light rays on it. We already assume that what happens in higher levels of the universe depends on what occurs in lower ones, right down to the subatomic particles, to light, and lower. That is upward action. What about downward action? The Brain acts downward on its parts by causing things to happen to its neurons, for instance. We can think of this as an analogy for the Universe with its series of levels, each of which has its specific entities and laws. It similarly interacts downward with each of its levels and parts. Imagine water flowing down through channels in a terraced garden. A single source at the top divides into several ducts and then divides and divides, passing through level after level of flowers, shrubs, and statues. So the Universe's interactions start at the top and flow down through its levels.

My memory of Pukeiti, an azalea and rhododendron garden in New Zealand's Pouakai mountain range, confers peace and inspiration. Shrubs stretch upward in glorious flower; others squat intricate and splendid. Paths wind to unknown dells. The garden's overall atmosphere carries to each plant within it. If the same tree grew on the curbside of a New York City freeway, with the same shape and flower, it would not be the same tree. The garden equips it with something different than does the freeway: serenity and excitement. The situation of the tree adds to the tree. The Universe's downward interactions, too, create a difference to its parts by conferring on each of them a wholeness of relationships with all other parts. A part by itself lacks this extra.

How does the Universe act downwardly? By what means could this happen? It's easy to state the fact that the world produces enough food for everyone to eat and drink what they need. But it's hard to break through the politics, beliefs, and self-interests to share the food around. Similarly, it's easy to say the Universe acts downwardly, but it's hard to describe a credible means by which the Universe could function this way. We need to, though. We need to know how the Universe acts downwardly so we can ask questions about it, so we can fix the idea if problems beset it, and so we can at least say when the idea will work and when not. We need to know how the Universe acts downwardly or else *downward action* is just a pretty term, maybe inspiring, but without concrete content.

Intuition whispers that the downward action of the Universe relates to its wholeness. Then if we specify the wholeness—thus saying how the Universe differs from the universe—out will pop the way its downward action operates.

Maybe Sperry's model of how the Brain or mind acts on the body can help us with the wholeness. Sperry's approach seems to be a good one, says the psychologist Aviel Goodman, because it draws together physical mechanisms and holistic activity, bridging the worlds of science and the experiencing self. But, he points out, it fails to deliver. Unfortunately, Sperry neglects to tell us how the whole (the Brain) acts on its parts (the neurons, for instance)—a serious flaw, according to Goodman. We should look at other holistic phenomena for clues as to how the downward action of the Universe works.

We looked at nonlocality in chapter 2. If you change the spin of a particle and observe what happens to its sibling, you'll see the sibling alters its spin at the same time as the other. This can also happen if the particles aren't siblings. Quantum theory only requires that they cohabit at some point in their past, and every particle satisfies this, because early in its life the universe squeezed all units of energy together. Thus nonlocality says that simultaneous events separated in space can correlate with each other though no known connection exists between them. Neither pulls the other's string or sends it a fax. Though it only affects quantum-level properties, nonlocality occurs at any scale from the subquantum to the supragalactic. And in theory, nonlocality happens everywhere, to events at opposite sides of the universe as well as to those in neighboring backyards.

One big happy family: each component of the universe correlates nonlocally with all others. In chapter 3, I called this "universal nonlocality." This phenomenon creates the wholeness of the Universe because it enables each of its parts to work with every other part. It produces an indivisible Universe. (I should clarify this with "helps to create the wholeness," because universal nonlocality is possibly only one way to describe the wholeness of the Universe. Nonlocality produces some of the wholeness. Out of the several types of wholeness, David Schindler of Notre Dame University points out, David Bohm employs only one: the interactions between the parts.) The capital-U Universe is the small-u universe viewed as a whole, thoroughly correlated through nonlocality. We might think of the Universe as a gigantic heater control where everything projects a button that, when pushed, instantly alters the setting of every other one.

We want to know how the Universe acts downwardly. The answer, as intuition suggested, lies in the wholeness: the Universe employs nonlocality to interact downwardly with its parts.

In *And the Trees Clap Their Hands*, Virginia Stem Owens explores nonlocality. For her, energy, the Spirit of the Divine, the underlying order, forms "by far the largest 'part' . . . of matter. . . . It is the Divine's life that flows through the arteries of the world, that seeps in the capillaries enclosing each quark, that sustains being at every moment." Further, "It is the Divine who thinks the whole, rounded thought of the universe. And as one thought, its nature, its total order, is indeed implicit."[3] Owens equates the Spirit of the Divine with the nonlocal unfolding of the underlying order's wholeness. She describes it as the "withinness," or the subjective experience of life and matter.

Universal nonlocality takes on the character of an all-knowing and everywhere-present deity that breaks free from the restrictions of time and space. This divine seems to correlate all events, every facet of our lives, in its embrace.

Yet I hesitate to equate the Divine with the Universe. The Divine as subuniverse unfolds everything, including nonlocal correlations, expressions of the underlying wholeness. The Universe therefore represents the unfolded order of our experience in a way that highlights the wholeness of the underlying or enfolded order. To speak of the Divine acting through nonlocality *approximates* talk of the Divine acting through unfolding. Thus we can suggest that the Divine relates or interacts downwardly with every part of the universe through the nonlocality of the Universe. In as far as nonlocal correlations are the unfolding of the subuniverse, so they are the means by which the Divine interacts with the universe.

Interacting with Specific Events

I raised an energy problem earlier in this chapter as a matter any explanation of specific actions of the Divine ought to solve. In a closed system, the energy level remains constant. But divine interactions create events inexplicable by physical laws, and these require or supply energy beyond that in the system. Impossible, we say. How does the nonlocality theory fare on this score?

Let us consider closed and open systems again. The inventors and manufacturers of refrigerators and polystyrene cups were originally unaware of the harmful effects of the fluorocarbons generated in the production and disposal of their products. People in southern lands contract more and more skin cancer because of the holes in the ozone layer that result from the use of these products in the Northern Hemisphere. Who would have thought that such helpful products as refrigerators and Styrofoam coffee cups would lead to global warming and, perhaps, the end of civilization and most forms of life on earth? Connections often surprise us.

Imagine if the United States government defined a "problem" simply as a disaster that happens within U.S. jurisdictions or something that threatens its strategic interests. A country such as Australia barely concerns the U.S., and its cancer dilemma occurs outside U.S. territory. So the Antarctic hole in the ozone layer poses no problem. . . . This definition of what forms a difficulty ignores the fact that the consumer mentality of the U.S. population and industry causes much of the threat to Australians and others.

You can't fully close off a system from the outside. The connections between the system and what lies outside it will always break the quarantine you hope to achieve. Nonlocality, for instance, will puncture the seal because it correlates the contents with every other part of the universe. Because the Divine interacts nonlocally with objects in the universe, the seal rules out much of the potential activity of the Divine with the system. When you label a system closed, you become an atheist.

What if we consider a system that is, from a physical point of view, approximately closed? We could attribute the neglected effects to experimental error and, for all intents and purposes, forget about them. An approximately closed system, however, encounters one of the problems of a closed system: it ignores the nonlocal correlations between the system and every object outside it. So it ignores possible interactions between it and the Divine. When you label a system approximately closed, you become—in the sense of divinity I'm developing here—an agnostic.

The idea of the Universe acting through nonlocality, therefore, avoids an energy problem. To isolate a system ignores valid nonlocal correlations with everything in the Universe, so it can only approximate. Then it ignores the relationships we seek to establish.

The interactions of the Divine with the universe are both holistic and deterministic; I characterize the way the Divine creates and sustains the universe as "holistic determinism." Determinism arises because the Divine unfolds the universe and all that happens in it, and thus determines what occurs (assuming the word *determines* applies to the Divine). Holism arises because the Divine creates and sustains the universe by unfolding a nonlocal Universe, all of whose parts relate. Everything the Divine executes, everything that happens—both the general and the specific—is holistic. Our brownie universe tastes different than the nuts and chocolate and flour do separately.

I suggest nonlocality as a means through which the Divine might interact with particular events in the universe. It respects the conservation of energy.

You may feel shortchanged. You may await a miracle, something new, unexpected, and unnatural that only the Divine could do. You want a finger from the sky, a lightning bolt that eradicates crime or disease or unhappiness forever. Then you'll feel satisfied. In the traditional sense of the word *miracle,* the nonlocality explanation flunks the test; it involves only ordinary procedures.

But I try to fathom the real Divine we believe in and the ways in which this divinity interacts with the universe and its parts. I want to understand what happens, not to fantasize. We can forget about events as miracles in the traditional sense. They lie outside the belief system of the regular Westerner.

From our perspective, what could the Divine do with nonlocality? Could the Divine send a couple to rescue two bewildered foreigners headed for Princeton Theological Seminary? The Divine unfolds miracles. Miracles are real. The meeting at the JFK terminal arose from a general action of the Divine (nonlocality) and specific decisions or inclinations by humans (the couple who drove us to Princeton). Maybe nonlocality can make anything possible, and whether it happens depends on other circumstances (a person's decisions or listening ability, for instance). All nonlocal activities present, like everything else, the Divine at work, but at work in a special way: namely, through the highly holistic phenomenon of nonlocality.

CHAPTER SEVEN

WHAT DOES THE DIVINE DO?

Picturing Divine Activity

"THE SUN RISES IN THE MORNING because otherwise we'd sleep too long and lag behind in our work. The Divine designed it that way because we must finish that work." I remember reading this in my childhood church's newsletter and thinking something like, "Something's fishy here. The sun rises because of the spin of the earth. The Divine designed the sun and the solar system for reasons other than to meet our needs; we evolved to sleep and work on a planet whose sun rises and sets every twenty-four hours." In the light of this science, the spiritual explanation struck me as senseless.

The sun "rises" in the morning because of our perceptions of inter-actions between the physical properties of the sun and the earth. As rou-tine work, the Divine carries out these properties in the ways physics describes. Unlike the church newsletter, this picture of how the Divine interacts with the universe runs along with science; the actions continue in the background and produce what science investigates and describes with laws. The Divine unfolds everything into existence at the big bang and throughout time.

Further, the Divine needn't go out of the divine way for beings like us. The Divine doesn't pamper us as if we are the apples of the divine eyes or punish us with the mandate to work because Eve and Adam ate the forbidden fruit. The picture of divine action that I'm presenting—the unfolding of the subuniverse—avoids personal questions about divine activities. We can leave aside why the Divine acts this way. Joshua commanded:

"Sun, stand still over Gibeon,
    and, moon, you also, over the Vale of Aijalon."
And the sun stood still, and the moon halted,
    till the people had vengeance on their enemies.[1]

The sun rises in the morning; but it also stood still for twenty-four hours when Joshua led his people against the Amorites. And, as with my airport rescue, some of these actions hold personal meaning: the Israelites believe the Divine fights for them. Believers think the Divine acts in general and behind-the-scene ways. Many also suppose that the Divine at times interacts directly with specific events in their lives, in manners that might break the laws of science.

Developing and Adhering to Belief Systems

The last chapter developed a means by which we can understand how the Divine can interact with individual events, but we have yet to discuss what the Divine would carry out with this ability. What do the nonlocal correlations of the Universe lead to in human experience? In what ways, when, and where might the Divine act in addition to the routine? Would nonlocality force the sun to rise early so that a Saint Belfry could see to feed bats at one o'clock each morning and then set so he could complete his eight hours' sleep?

Let us look more closely at beliefs. Anthropologist A. P. Elkin writes:

When a Djauan tribesman in south-west Arnhem Land, Northern Territory of Australia, said he has just seen a mimi—a being in their spirit world—represented in rock-shelter galleries, he was reporting what was an actual experience. That I could not see the mimi in spite of his guidance was irrelevant. I had not imbibed the belief and the traditional context, and so did not have eyes to see. . . .

In 1924 a full-blood Aborigine from the mouth of the Murray River, South Australia, was staying in my home in the Hunter River Valley, New South Wales. Walking by a billabong in the paddock, he said that if a pelican settled on it, he would know it brought him bad news from his people, and he would have to return to them at once. He did not ask whether pelicans ever alighted on my billabong. He took for granted that his *ngaitye,* the "friend" of his clan, would appear in case of trouble. We might dismiss this expectation as baseless, but to him it was a tradition of totemic experience, of a mystic relationship between [people] and nature (species and phenomena), into which he was born. My Narrinyeri guest expressed his accepted belief which itself arose out of, and was nurtured by, a believed and quite credible experience of past generations.[2]

The belief systems of the Aborigines work for them in their original cultural and geographical settings. How can this happen? How do belief systems "work" so that their believers obtain the goals the system allows?

The study of belief systems and their effectiveness helps us understand what the Divine can carry out in our lives. James Borhek and Richard Curtis define a *belief system* as "a set of related ideas (learned and shared), which has some permanence, and to which individuals and/or groups exhibit some commitment."[3] Their book, *A Sociology of Belief*, describes beliefs as cultural phenomena arranged in systems and carried by groups of people, the believers. They continue by outlining the elements of a belief system, adding that the believers might fail to recognize them and that a belief system might exclude some of them. Elements include

- values or goals that determine the good and valuable in behavior;
- criteria for determining the validity of a statement;
- a logic or language, often implicit, that prescribes the rules for relating one belief to another within the system;
- a perspective that relates it to other things, groups, and belief systems and describes the group plus the place of each individual within it;
- beliefs that the believer considers primary;
- prescriptions and proscriptions, saying what one can and can't do; and
- a "technology," the means for obtaining the desired goals of the belief system.

The Aboriginal belief system desires close relationships among its people, and the pelican acts as a means to achieve this. For the Narrinyeri guest of Elkin, the pelican operated as a technology.

We develop and adhere to belief systems to create meaning. They define what constitutes understanding: the landing of the pelican or the ultimacy of the physical universe. Whether you exist as an Australian Aborigine or an American atheist, one or more belief systems hold you. To think and understand, to create sense of the world and of experience, establishes humanness. Without eyes, we can't see.

And the believing lens through which we peer at the universe adheres inseparably to the universe itself. We believe the truth of what we believe.

The belief system of Elkin's guest worked for him. The couple befriending my wife and me at the JFK airport justified the trust I held in the facilitator of my life. My belief system works for me, too.

When you flick a switch and the room fills with light, you experience the influence of your belief system. You believe in the power of technology and science to manipulate the physical world to obtain from it what you want. You believe that the theories of science approximate the real world, so that when you operate according to them, you can predict the effects in the world. You act accordingly, and the light turns on. Despite the light system using arranged materials and confirmed elements, its success still reflects the power of belief.

If you lay in a coronary care unit and someone prayed for you daily, studies show you might face fewer life-threatening events and complications than if no one prayed for you. If you embrace a spiritual commitment, you also possibly experience satisfaction with your life, plus feel better and happier. Most likely you enjoy mental and physical health. You will probably register a lower blood pressure and a lower chance of suffering from cardiovascular disease. You may live longer. According to studies detailed in "The Forgotten Factor in Physical and Mental Health: What Does the Research Show?" by David Larson and Susan Larson, people's belief systems can exert a positive effect on their health. The diversity of belief systems means they sometimes contradict each other. But they all work to some extent.

Approaching Others' Belief Systems

Do we notice knee-jerk reactions to beliefs other than our own? "Theirs don't work." "They can succeed a small amount only when they parallel science or common sense." "Believing moves the patient into a more healthful frame of mind." "The Aborigine's family and the pelican—that's just a fluke." We unwittingly think ourselves culturally superior and dismiss the belief systems of others, including those within our society—Scientology and communism, for instance—whose beliefs differ from what we hold. Their beliefs threaten us because we understand the world with our own beliefs, and we want the power we sense this gives us. It makes us feel safe and helps us pull away from hints of chaos.

But I say, Rule out fluke as a response. Approach the belief systems of other people seriously. That the world makes sense to them, and that it operates in the way they think it does, isn't a matter of coincidence. It isn't a coincidence that when a pelican lands on a billabong, something wrong has happened back home. It isn't a coincidence that the couple met me at JFK. It isn't a coincidence that people recover faster if they follow a spiritual life and believe a higher power will cure them. It isn't a coincidence that light fills the room when you flick the switch.

Belief systems succeed neither by fluke nor by approximating scientific truth. *Our* system works not because it lies closer than *theirs* to how the universe "really is." Every belief system can think that way (whatever they accept as the "really is"), however. Perhaps the breadth and depth of the universe encompass enough room for many versions of "really is," for many truths. This suggests that our alternatives are ours and so are inappropriate as the standard for all belief systems. Aborigines think differently. Their equivalent explanation about the pelican probably involves the Dreamtime ancestors of people, spirits, and places; they know the reality of the correlations between the Dreamtime and what they experience. To explain the functioning of belief systems—Aboriginal or scientific, contradictory and diverse—we should avoid prejudice against any of them. So how might we explain the success of belief systems?

Explaining the Success of Belief Systems

Roger Penrose and University of Arizona anesthesiologist Stuart Hameroff claim a specific idea of how consciousness emerges from the brain. In particular, consciousness enters existence when nonlocal processing in neural structures called microtubules reaches a critical level. Several other neuroscientists also associate quantum physical activity in the brain with consciousness.

If the Penrose-Hameroff or a similar theory stands up to critical examination, thinking involves nonlocality, a weird quantum phenomenon. But nonlocality automatically means correlations with the Universe. Strange though it may seem, thinking interacts with the Universe.

Belief systems underlie what we think. They underlie our minds. They provide cohesion to experience. A close correlation exists between thinking and belief systems. So I offer this hypothesis, open to investigation: the thinking of believers correlates nonlocally with the Universe, and this allows belief systems to interact with what happens in it. A belief system may represent a particular alignment within the microtubules of the brain. When the believer thinks, the nonlocal quantum events in these microtubules interact nonlocally with and so affect the universe outside the brain. Events outside the brain also correlate nonlocally with those inside.

But many of my wishes stay unfulfilled. However much cross-country skiers like myself long for ideal weather, it seldom arrives. Belief systems may fail to work, and the nonlocality explanation for their effectiveness must allow for such setbacks. In what circumstances will a system of beliefs succeed and when will it not?

• Skiing lies outside the core of my belief system. Maybe you must hold ardently to and follow the belief system for it to deliver.

• Whatever structure skiing represents in the brain may malfit what microtubules can achieve for beliefs. The physical constraints of microtubules may affect what a person can believe and thus the potency of the belief system. Because of brain structure, maybe not everything goes, at least not every belief.

• Believing leads to quantum effects because nonlocality registers at the quantum level. Will these quantum consequences translate to what believers can observe—skiable weather, for instance? The relationship between the micro and the macro worlds may also explain why some belief systems don't bear fruit in places and why others flower like crazy in the same circumstances.

• A belief system that demands people quit eating, drinking, and sexing will show complete disinterest in cross-country skiing because fun feels foreign to it, and because its "scriptures" neglect to mention any form of the sport. So skiers should hardly expect the winter weather to improve after pleading their case through it. This suggests that a belief system may work poorly for the particular portion of the universe that we seek to change because the system focuses on other matters. It may only work well for us if it focuses on the area of the universe in which we want change to happen.

What believers believe will happen may happen under the circumstances their belief system prescribes. This can occur because belief systems, through the mind and brain, interact nonlocally with the universe. Under the right conditions, the Universe—the universe nonlocally correlated—will perform as believers expect. Or to rephrase it, the belief system, if successful—which means if its adherents survive well—works in a way that coheres with the universe and its nonlocal interactions.

Under the right conditions, the Divine, the unfolder of the universe, will perform as believers expect through their belief system.

Believing in Creativity and Freedom

To pose the question more pointedly, I should ask: What individual events in the life of the modern Westerner happen as a result of our belief system? Our interest, after all, lies in ourselves. What specific nonlocal interactions does the Universe play a part in and that we experience? What specific events do we believe in and that stump science? Let's look at some examples.

> • I once saw a man pushing his daughter in a stroller up a hill. A car driving up the hill gradually veered across the opposing lane and over the curb. Thanks to quick reflexes on the father's part, the car just missed him and his daughter. It smashed into a concrete post and stopped. The father couldn't see the driver. Quick reactions again, he set the brakes on the stroller and ran to the car. The doors were jammed shut. While bystanders stood around stunned, he told an onlooker to phone for help and smashed his way in. He feared fire and could see a body on the floor.
>
> Why did the father choose to act quickly to help the driver?
>
> • To solve a problem you find when writing, instructor and journalist Donald Murray suggests fast writing: put down whatever comes to your mind, with no concern for grammar or spelling. The solution will work itself out, will emerge from your words.
>
> Where does the answer come from?

Creativity is the common theme of these two incidents. The second tells how writers might act to solve an intellectual problem and how they can assist the creative insight to emerge. The first example involves a father's freewill action to help someone whose car nearly hit him and his child. In that he chose to act against his fear and shock, his choice involved creativity.

Can we predict the lyrics and tunes of the next show Andrew Lloyd Weber will create? Can we predict the next theory that will revolutionize biology? The questions answer themselves. Our Western belief system says that all sorts of scientifically explicable events happen. But which ideas creative scientists and technologists advance lies outside scientific prediction.

Our culture believes deeply in creativity and freedom.

When the car headed his way, I imagine the father drew on physics, his driving history, knowledge of human nature, experience of the stroller in fast movement, his concern for others, fear of getting involved, love for his daughter, the way he had strapped her into the stroller, his fear for his life, his awareness of what he could do under pressure, a sense of trust, courage, and more. Nonlocality forwards such elements to us, elements from both inside and outside the brain. Everyone had access to what impinged on his mind when the car crossed the curb. Creativity includes drawing on a wide range of elements and assembling them in a unique way.

Creativity and insight exemplify individual interactions with the Universe, examples of engagement arising from our system of belief. Nonlocality gathers the elements for creative insight. But nonlocality depicts the holistic interactions of the Divine with the parts of the universe. Therefore, we can say the Divine acts via nonlocality to offer the ingredients for unique thought.

I'm no Mozart or Penrose. It's one thing to have the elements available nonlocally; it's another to engage them uniquely. The second aspect of creativity now appears. Rather than drawing on all of the correlations, we filter them. What supplies the unique pulling together? The father chose how to act when the car crossed the road. With all the information available to him, his brain decided without a second thought. Biology provides the ability to fit them into a pattern. The choice emerges from the individual's brain influenced by upbringing, education, and culture. Thus people draw on the elements in different ways; some—the Mozarts and Penroses— show more originality than others.

The Divine supplies and we choose. Creativity and novel thought require an interaction between the Divine's and our endeavors.

The woman who invited us to ride with her and her husband to Princeton knew something. She may have overheard our destination in our attempts at phoning, and she must have sensed our distress. She may also have intuited other information about us. Either way involved nonlocality and hence the Divine at work. Her contribution to the miracle? She was a very sympathetic person and picked up our cues.

The other factor in the JFK miracle concerns our belief system. The Divine acting nonlocally produces the ingredients for our choices. A system of belief guides the believer in how to use this information. Caring for others is something we believe in, and so the Divine connects us with details and we, with the guidance of our belief system, choose to act in a caring way or not. The salvation at JFK fits right in with how the Divine produces creative insight and the effects foreseen by the belief system.

Such effects are downward interactions by the Divine because of the holistic nature of nonlocality. The Divine acts through the Universe's nonlocal correlations. The effects are also specific because the belief system supports such activities: under the right conditions, the unfolder of the universe, the Divine, will perform as believers expect through their belief system.

CHAPTER EIGHT

THE REAL DIVINE

Emphasizing Divine Action

THE DIVINE IS THE UNDERLYING SUBUNIVERSE, the patron of existence and logic. The Divine as subuniverse creates the universe at the big bang. The initial creation also involves continuous creation, as the action of the subuniverse created time then too. The Divine is that which does all, all changes and nonchanges, producing everything (both subjective and objective), sustaining it in existence moment by moment. Every event is a divine action. We describe some of these activities of the Divine with scientific laws, which means that scientific theories describe the subuniverse Divine at work. The subuniverse, the Divine, pulses as the heart of the universe.

The music of the Boston Symphony Orchestra under Seiji Ozawa held me and the rest of the audience on a thread out beyond the rail. We hung there suspended in a trance. It then lowered us back into our seats, where we smiled the broadest grins our faces could stretch. Likewise, the Divine performs the universe.

Because most of us doubt like Thomas, we lack encouragement to believe unless we see the marks of the Divine at work—which now hide, tricky and ambiguous. This spells death for spiritual belief. So I reject the possibility that the Divine interacts with the universe only from time to time or at particular places, whose origin obscurity cloaks. I want to emphasize the relevance of the Divine for the ordinary life of modern Westerners.

Is the subuniverse Divine real? Does the Divine–universe relationship that we pursue provide a divinity we readily experience?

Interacting by Means of Down Trickling

The name of an Oxford biochemist and Christian cleric, Arthur Peacocke, frequently pops up when scholars discuss how the Divine interacts with the universe. His scheme matches mine except his divinity exists independent of the Universe. His Divine interacts downward with it, which then interacts downward with its various levels. We may ask, as I did of my proposal, how Peacocke's divinity works downward. By what means does Peacocke's Divine interact with the Universe and the Universe with its parts?

Cambridge physicist and Christian cleric John Polkinghorne is often cited when scholars discuss how these different levels relate to each other. Though he finds the difficulties "mostly too hard for current knowledge,"[1] he provides an answer that Peacocke also picks up on: information transfer.

A rock falls in your path just around the corner. A message flashes in your consciousness to act suddenly and so avoid the danger. The Divine communicates information to the Universe, and like a repeater station, it passes the information on to us with physical interactions, such as encouraging thoughts and feelings in our minds.

Information lies at the core of Peacocke's image of the Divine–universe relationship. Information bridges the levels. With it, the Divine influences events. It accounts for the Divine's activity in the universe and respects the explanations of science.

Information seeps down to the Universe and then filters to the various levels and parts of the universe. Thus I call this version of divine interaction with the universe "down trickling."

Trickle-down theory infused U.S. politics and economics in the 1980s. The economically depressed will ultimately benefit, the policy stated, when government aids big business through loans and tax abatement. Large tax cuts place money in the hands of corporations and affluent individuals who invest in job-creating industries, and so wealth at the top gradually trickles down to those at the bottom. During the 1980s, though, the riches of the rich grew exponentially while the impoverished lost out. The number of poor people expanded (an additional seven million people in the United States fell below the poverty line), and the middle class shrank. Many at the bottom and in the middle still wait for money to trickle down.

A similar occurrence can happen with governments. The more levels of administration and management proliferate, the less efficiently government operates, and the less likely it performs as the people and lawmakers want it to. Good intentions may fail to trickle down through many levels.

Children play a game in which the first person whispers something in the second person's ear ("Simple Simon met a pie man"), who whispers it to the third person, and so on around the circle. The last child says out loud what she or he heard: "Swishing swishes meat purple pigs."

The interaction of the Divine with the Universe similarly dissipates as it trickles down the levels to us and beyond. The farther apart two levels lie, the fewer trickles from the higher one drip onto the lower. The down-trickling theory of divine action means the effects of the Divine on the human world appear minimal, barely noticeable, nearly invisible. In the regular course of nature, they become unobservable.

This avoids identifying specific events as the Divine at work in people's lives; we should expect never to witness miracles. The Divine doesn't exercise this sort of power. Thus we can forgo explaining why the Divine stopped this suffering but allows that misery. It also provides the independence of the universe from the Divine, protecting our freedom from divine control. Neither do we face with it the problem of the Divine's meddling in the universe; we escape the haunting "energy from nothing" problem. The down-trickling theory thus offers advantages.

Knowing the Divine Reality

We know the reality of something through its effects on us or our observed effects on it. The reality of the Divine ties closely with what the Divine does, so for the Divine to be real, we should notice a lot of what this divinity carries out. We should observe obvious divine actions. The Divine must markedly affect most people's lives and the universe we experience. But a divinity that interacts with the universe through information transfer affects the world of our common experience very little. It is thus an inadequate divinity. It is weak. The reality of the Divine in our lives suffers when the Divine interacts through down trickling.

What does the Divine do? Fundamentalists view the Bible in a particular, literal way. It speaks to them about the nature of the Divine, and the answer to a question lies before them written on its pages. Orthodox Roman Catholics look to church tradition. It protects the vault of truth because they believe the church speaks the one and only infallible voice of the Divine. Those who roll in emotional experience might follow a charismatic or a new age line. The truth, they believe, must carry away certain of their feelings.

What does the subuniverse Divine do? The answer depends on how you think the Divine might act. Experience of the Divine ties to your beliefs about the Divine. Fundamentalists, orthodox Roman Catholics, Charismatics, and New Agers observe the Divine at work where their beliefs say they'll catch the evidence. The reality of divine deeds attaches to their recognizability, to your perspective. So what would render the Divine real to most people? Something that the beliefs of most of us in the modern West would define as real. What might that be? Our outlook emphasizes the here and now, the universe and the world of human interaction and experience. To produce the reality of the subuniverse Divine, divine actions should perform on the stage of the secular.

Sociologist Richard Moran proposes a new economic approach,

> The "percolate-up" theory: The best way to stimulate the economy and increase the standard of living is to put more money into the hands of the poor. Countless studies have shown that poor people spend a greater proportion of their income than do rich people. More spending requires more production, more production requires more employment and this in turn will generate more demand, stimulate more production and create even more employment.
>
> Consumers, not producers, will create the energy for economic growth. Thus the benefits of putting more money into the hands of the poor will gradually percolate up through the middle class and beyond.[2]

For our system of spiritual ideas, we start at the bottom, the basics, what our feet stand on. Every event is the Divine at work. The Divine is the origin and source for the existence of everything, their sustainer moment by moment: the falling of a ball, the evolution of species, the universe's big bang. What we perceive as simple regularities, as complex phenomena based on simple regularities, or as unique phenomena based on the regularities—and a lot more of which we are at present unaware, and I imagine some of which we'll never be aware: the Divine unfolds everything. The universe resembles the doll of a ventriloquist. When watching, we feel life in the puppet without awareness of the performer's part; it carries us away. The Divine performs the universe.

Because of our secular-scientific belief in the reality of the universe, nothing affects the universe more observably than the Divine. Because of the reality of the universe around us and all that happens in it, both regular and unique, the actions of the Divine—and hence the Divine—are real. Any person who approaches the natural world as the fundamental reality will find this true of this natural divinity. The here and now roots the totally involved Divine. The Divine, therefore, inhabits the everyday experience of most people; the subuniverse Divine is the divinity of at least the Western belief system, maybe of others as well. The Divine really acts because the universe really exists.

Living assumes this divinity (not always consciously), so the Divine unifies life, our lives.

Examining the Universe to Learn about the Divine

Consider a word such as *attraction*—meaning physical attraction but not sexual attraction. Two bodies attract each other under gravity. The Divine unfolds the gravitational events as if the bodies attract each other. The Divine produces the universe according to the regularities of natural laws. Divine acts subject bodies to something we perceive classically as an inverse-square law. Does the Divine bear a mathematical nature because the universe follows mathematical regularities? Forgetting that we built logic into the idea of the subuniverse and hence of the Divine, the question becomes empirical. The Divine's reality helps us know about the Divine.

Recent research in particle physics shows that "hair"—made from fields that give rise to subatomic particles such as pions, or W and Z bosons—protrudes from some black holes. Perhaps the Divine favors hair. Perhaps the Divine is quite hirsute, long white beard and all. The universe behaves in certain ways, and since the Divine unfolds them, they say something about the Divine. The Divine acts like this, so the Divine must be like this. The pot reflects the mind and perhaps the fingers of the potter.

The universe reflects the nature of the divine subuniverse, so you can examine it to learn about the Divinity. That's if you're open. The Divine, like the horse, is real, unlike the unicorn of imagination. You can study the horse and the Divine, not the unicorn.

If the universe presents basic properties other than logic and fruitfulness, and if we accept them as divine attributes, we need to show how they emerge from our model of the Divine. If we attribute other properties to the Divine, these too should appear as basic features of the unfolding universe. That same universe founds our ideas of the Divine. Only investigation of the universe can justify a belief, though the idea may arise from any source. The empirical method, a treasure of our culture, provides a portion of the route to knowledge of the Divine. We ask the universe scientific questions about the Divine.

The ventriloquist determines every movement of and sound from the puppet. Does the Divine determine everything we and every other part of the universe do? Are we genuinely free? The subuniverse approach invites us to cover topics such as evil and suffering, the independence of universe and the Divine, pantheism, and freedom.

CHAPTER NINE

MYSTERY

Upholding Immanence and Tradition

TRADITION POINTS TO THE GROSS INADEQUACY of our pictures of the
Divine, reminding us that our spiritual experiences fall far short of who
the Divine is. This idea of transcendence constitutes one side of the story.
But if we hope to catch the reality of spiritual encounter, our model must
also closely tie with everyday experience. Custom calls this approach of so
anchoring our image "immanence." Immanence and transcendence conflict with one another, so doctrine embraces them in a tension, as if separating two fighting children. In my system of spiritual ideas, I assign
immanence, or everyday experience, considerable importance. Do I
thereby displace transcendence?

The Dutch philosopher Baruch Spinoza considered mind and body
(or ideas and the universe) as two aspects of the same thing, which he
would alternately call "God" and "Nature," *God* being *Nature* in its full-
ness. His philosophy upholds both transcendence and immanence without
a dualism. But the religious and political authorities of his day regarded it
as blasphemous, accusing him of the heresy of pantheism, the belief that
everything is divine. Does the system of spiritual ideas I propose succumb
to Spinoza's heresy, too?

In a discussion on David Bohm's thought, Robert Russell suggests the
Divine is what I call the "subuniverse." He asks if tradition would classify
this image as pantheist because of the intimate connection between the
universe and the Divine. No, he answers. Bohm's ideas repel pantheism
because they point to transcendent features of nature. The subuniverse surpasses the order of our experience, for instance. The Divine transcends

everything. On balance, Russell concludes, this image subscribes to a pan*ent*heist view in which the Divine envelops and exceeds the universe. Does this satisfy the needs of tradition?

The subuniverse is a gigantic whole that produces the universe and accounts for everything that happens in it; as a self-contained sack around the embryo universe, it creates and supports existence. Talk of the whole suggests the Divine, writes Ted Peters. But does the subuniverse coincide with the Divine? Orthodox Christians, Peters continues, would disclaim its holiness if that meant "sacred." They could only think it holy in the sense that *holy* and *whole* emerge from the same root word.

Affirming the Divinity of the Subuniverse in the Face of Christian Doctrine

Peters believes that to affirm the divinity of the subuniverse denies two Christian doctrines. First, it fails to distinguish between the Divine and the universe. Traditional belief requires such a difference, similar to the distinction between an architect and the building she or he designs. Arthur Peacocke would support this conclusion of Peters. The toughest problem for my spiritual proposal, he once told me, is the idea of transcendence. Many hard-won ideas in the orthodox portfolio—for example, the independence of the universe and humans from the Divine, of the Divine from the universe, free will, the moral challenge that always stands before us—depend on a firmly held sense of divine transcendence. Other critics may want to safeguard the Divine they experience as something quite different from themselves.

Second, Peters thinks that equating the Divine with the subuniverse counters the doctrine that the universe depends entirely on the Divine's gift of existence. Our alternative aligns creativity with the subuniverse, a material entity, while Peters emphatically distinguishes it from the Divine. He and many others want to avoid the charge of pantheism, the black hole in their universe.

At its end, the universe will finally collapse into the big crunch or slide out in an eternity of cold death. If the Divine is the subuniverse, these events will significantly affect—perhaps even kill off—the Divine. That's impossible according to John Polkinghorne. Something must prevent such dramatic happenings. Perhaps the Divine will unleash naked power to stop such "changes and chances of this fleeting world" from sweeping the Divine along with them. But this would block the universe from following its inbuilt laws; the universe would lose its freedom. Thus Polkinghorne rules out dramatic intervention. The only option he sees grants the Divine considerable independence from the universe: the depths of the Divine depend on nothing created, nothing in the universe. Polkinghorne therefore rejects systems of spiritual ideas that tie the Divine closely to the universe—as with our subuniverse theory—because then the Divine becomes overly vulnerable to what happens in it.

Understanding Transcendence

Most people of Muslim, Jewish, and Christian backgrounds would accept these arguments for tradition as reasonable. They would require transcendence in their image of the Divine. This then raises a question central for our system of spiritual ideas: If I focus on wholeness and define the underlying subuniverse as the Divine, how might I understand transcendence? I reject the easy option that nothing we can know or experience matches the Divine. To say the Divine transcends everything completely does manage the difficulty but falls short of wholeness. So where should we look for an answer? When we find one, will it satisfy the needs of tradition?

The Christian faith believes in the Divine as being at the top as Father, the mysterious creator who transcends everything. At the bottom, as it were, the Divine as Spirit permeates the universe and human experience, radiating and immanent throughout the complex system of reality. Many spiritual traditions bestow on the Divine this twin role of immanence and transcendence. The Western psyche internalizes its dual emphasis.

This book's system of spiritual ideas speaks to it as well. The Divine begins at the top as creator. As the subuniverse, the Divine brings about the existence of the universe and its parts at every moment. Further, as every object builds from the most elemental parts of matter—leptons, quarks, and bosons—so all things and experiences start from the Divine at the bottom. As we partly understand anything by the way its parts behave—the biochemistry of our bodies, for instance—so we partly understand anything, including life, by its emergence from the Divine. The Divine begins at the bottom and seeps back through to the top. So my proposed system of spiritual ideas also pushes immanence. It supports the tradition of immanence and transcendence.

Refocusing on Mystery

I know little about how my university's administration works, and so I grant it is a mystery, as I do with anything unfamiliar and powerful. The word *mystery* catches my eye and my gut more than does *transcendence*. It feels right. *Transcendence* intellectualizes *mystery; mystery* underlies *transcendence*. Transcendence, though, may merely cipher our inability to comprehend; in this case, either we assume this or it reflects the inadequacy of our attempts to understand it. I have rejected the first assumption about the Divine because I believe we can comprehend something of the Divine. The mystery therefore reflects our ignorance of the Divine, ignorance we could satisfy by finding out more. So, I refocus the discussion of transcendence on mystery and ask how we might understand divine mystery.

Transcendence and immanence both exist in our system of spiritual ideas. They also intertwine in mystery. The interactions of the Divine with the universe look obvious and readily at hand as the discussion on the Divine's reality points out. Transcending with radical immanence, the Divine continuously brings about each event in the universe of our experience. Everything bears the mark of the Divine, is of the Divine. Mystery emerges from intimate involvement.

Mystery also emerges from wholeness. The subuniverse contains the past, present, and future of the universe, all its potential, and anything imperceptible to the foggiest margins of our knowing. Nothing could be more mysterious than the subuniverse Divine.

Just when I think I understand people around me, something they do or say baffles me. The whole person exceeds those parts I comprehend. Similarly, if we equate the properties of a simple and mechanical nature with those of the Divine, we lose the mystery and the Divine. But if we recognize the Divine as the subuniverse that unfolds our universe and every other one and so surpasses everything, including the capital-U Universe, then we confront mystery. The Divine exceeding the constituents of the universe is the most mysterious thing of all.

What if we come to understand the wholeness, the unfolding, or the potential through some future insight of human inventiveness? Does mystery then disappear? More to the point, does knowledge remove mystery?

We see the subuniverse through glimpses of its shadow, the world of our experience. At each sighting, some of it lies out of view. It unfolds only in part. What we know and will know of reality fails to exhaust it; we can never comprehend everything. Scientific and other knowledge grasp some of reality's significance, but they only partly express what they profess to cover. Nor can we decide, imagine, or sense by intuition how far reality eclipses our comprehension. Just when physicists think they have nailed the basic element of matter down to the top quark, evidence suggests it too comprises more fundamental particles. In principle, we can never know the universe in full. Bohm calls this "the endless depth of reality": the subuniverse and its universe elude us and so appear endless. In his words, nature possesses a qualitative infinity. This inexhaustible depth of nature, its unfathomableness, opens us to the Divine's mystery.

I can't know all about you. Your subconscious has closed off parts of your psyche and your past from others (as well as from yourself), so however much psychiatry and psychology delve into your mind, it remains inexhaustible. Knowledge arises only through metaphor, analogy, and model building, none of which corresponds completely to the object known, thus leaving discrepancies. The endless depth applies to the universe and to anything in it. Every object and process relates to all others and continuously changes in its unfoldings, rendering any freeze artificial. Each thing owns infinitely many sides and we experience only a finite number. No matter what we know of a thing, it exceeds our knowledge in ways that tell of mystery. The unexpected, or the edges, always lie in wait. Even if we know more about something in the future, it will still mysteriously exceed our knowledge. I call this inexhaustible depth to the reality of anything the "gap in knowing." The inspiring heights to which humans rise and the corrupt depths into which humans fall constantly remind us of it.

Acknowledging the Gap of Knowing

Yet can we know anything? Consider Alice:

> The rabbit-hole went straight on like a tunnel for some way, and then dipped suddenly down, so suddenly that Alice had not a moment to think about stopping herself before she found herself falling down a very deep well.
>
> Either the well was very deep, or she fell very slowly, for she had plenty of time as she went down to look about her, and to wonder what was going to happen next. First, she tried to look down and make out what she was coming to, but it was too dark to see anything; then she

looked at the sides of the well and noticed that they were filled with cupboards and book-shelves: here and there she saw maps and pictures hung upon pegs. . . .

Down, down, down. Would the fall *never* come to an end? "I wonder how many miles I've fallen by this time?" she said aloud. "I must be getting somewhere near the center of the earth. Let me see: that would be four thousand miles down, I think . . . yes, that's about the right distance. . . ."

Down, down, down. . . . She felt that she was dozing off, and had just begun to dream that she was walking hand in hand with [the cat] Dinah, and saying to her very earnestly, "Now, Dinah, tell me the truth: did you ever eat a bat?" when suddenly, thump! thump! down she came upon a heap of sticks and dry leaves, and the fall was over.[1]

Must we keep falling through the unknowable, or does a shelf jut out so we can save ourselves? Bohm notes that many scientific theories predict rightly some of the time. Approximate descriptions do exist, and reality retains some stability. We can know things despite the gap in knowing.

Theories predict correctly only some of the time. They build on ideas valid in a restricted setting and under specific conditions; each develops only a few insights, a narrow unfolding of the subuniverse. If we shone a light down Alice's well, we would see to a certain depth, and we would observe books and pictures protruding from its edges. Similarly, a theory penetrates the open and unknown—but only part of the way down and in a few directions.

Alice landed on a pile of sticks and leaves on a solid floor. Shelves of temporary certainty jut out into the endless depth of reality, the well of the unknown. We linger on a ledge of knowledge that works. If we land on one that fails to work, we try to patch it up or search for another. If it keeps failing, we look for another more adequate ledge. That's human nature. We should search and not forever fiddle with dubious or unclear parts of a theory, the places where reality avoids its grasp, because confusion might always surround it. Bohm thinks that puzzles will in this way point us toward new and more adequate insights: from Newtonian physics, then to relativity and the usual quantum physics, and now to his theory. This constant search in the face of perplexities undergirds the empirical or scientific method.

We may fear that the gap in knowing—the inexhaustible depth to the reality of anything—will land us on the floor of relativism, where all ideas possess equal truth. We needn't worry. New insights, each with its own limits, drive an area of knowledge toward something more truthful, more insightful. Science doesn't march toward a fixed set of facts that fully represents the universe. For Bohm, it hops from one limited theory to another restricted but more useful explanation.

In U.S. politics, responsibility hops from one limited seat of power to another restricted but powerful body. When the president signs an accord with a foreign government, he usually presents it to Congress for approval. The president represents, but is not, the government of the United States. Similarly, even if I were the world's greatest electronics engineer, what I would know of a computer wouldn't be the computer; it would only represent the computer. Because representations contrast in this way with the items represented, the gap in knowing continues to widen. Knowledge of something isn't the something; comprehension only exists in our brain. All knowledge bears this mark.

What I understand of the computer or anything else entwines with my experience. I may think I grasp the truth of the computer's operation, but this knowledge still involves some of who I am. What we know ties so tightly to who we are that our minds can't escape our selves; we can't stand outside ourselves and look at something from a point of view external to our experience. We can't climb out of ourselves to check out whether what we know is perfectly true or not. The gap in knowing widens further still.

Partly because experience of the universe provides our spiritual knowledge and partly because knowledge is knowledge whatever its subject, the same gap in knowing applies to what we know of the Divine. Because of our limitations, we can't know everything about the Divine. A lot about divinity will elude us forever. The Divine possesses infinite depth. Our understandings merely glimpse the unknown that forms both reality and the Divine, and they both slip through our concepts. Some of our understandings of the Divine may fit perfectly and some may miss the mark, but we'll never find out either way. And we can't know whether what we do know of the Divine is completely true. This realization allows for both mystery and knowledge.

Deciphering Differences

This realization also avoids both the principle that we can know nothing of the Divine and the idea that we can know everything about the Divine. Some claim a uniqueness for the term *the Divine* because we can know nothing about its referent. They think the Divine transcends the universe absolutely, unlike anything else. But the gap in knowing applies to all knowledge and names; that is how we employ words. *The Divine* isn't unique, because the gap in knowing applies to it. Citizens of the United States claim that red, white, and blue distinguish their flag from all others, but the flags of the British, French, and Dutch are also red, white, and blue. Thus the name "the Red, White, and Blue" clouds the nature of the contrast between the United States and other countries. The difference between flags means little, but the difference between the opportunities in different countries means a lot. Absolute transcendence clouds the contrast between the Divine and the universe. It says by fiat that a total difference exists, whereas the better—and more difficult—task deciphers the actual differences between them.

What is the difference? To paint a picture of the universe, we can start with what we know of it and add what we know that we don't know of it. This begins to fill out the description. Where does Alice go once she lands on the sticks and leaves? We can manipulate known things to some degree, and we can search for the unknowns to better our understanding of the universe. But there are also unknowns that we don't know, of which we are unaware. Maybe a secret passage led from the room Alice landed in to a room filled with treasure, but she didn't know about it or even suspect it existed. The Divine unfolds three categories: what we know, what we know that we don't know, and what we don't know that we don't know. All of these emerge from the Divine and comprise part of the mystery of the Divine. The Divine's mystery relates closely to the mystery of all things, differing by degree because it exceeds that of the universe of parts.

Mystery as depicted in the gap of knowing emphasizes the cognitive. It also anchors spiritual experience: some of us feel and sense something more within ourselves and our lives. We feel and sense an otherness that also connects with us closely. The system of spiritual ideas we develop expects a wealth of such spiritual encounter. In it we experience "the other." It is also experience of the material, in a larger sense of *material* than the phrase "nothing but matter" usually connotes; mystery permeates the material. Traditionalists would emphasize the spiritual as from another realm. In our system of thought, the spiritual—though natural and not supernatural—does differ from us. We understand a part of the Divine through scientific and rational means, but because of the gap in knowing, mystery will always engulf it.

The spiritual tradition in which most Westerners grew up nonetheless insists on absolute transcendence and does so for a reason. Does our model satisfy that purpose?

- It allows for encounters with "the other."
- The Divine eludes the grip of our knowledge and lies out of our control. We can't forge the Divine completely in our own images. Thus a morality embedded in our experience or knowledge of the Divine at least in part derives from outside our selfish wishes.
- Mystery speaks to the transcendence of the Divine over the universe of our experience, thus ruling out pantheism.
- The subuniverse Divine produces this universe and probably many other universes as well. Others will continue after the demise of ours. Thus the fate of the Divine needn't hinge on that of our universe; the Divine will continue if our universe dies.

Our model of the Divine satisfies what tradition intends by transcendence.

This model avoids jamming reality into a hard and square metal box that only relates to what we humans can control. We can't and won't control or know everything; science will continue. The emphasis on science in our system of spiritual ideas reduces nothing to mechanical principles. An infinite will always lie beyond us, waiting for us to venture into it. We enter it, we find out more, we enrich ourselves, we endanger ourselves, we excite ourselves, but we'll always see more.

The mystery and the key to understanding lie in this universe, the place where we "live and move and have our being." They thrive in the material world we experience, because here we encounter the mystery and the key to understanding. This ties the Divine into the ordinary of life. It involves the Divine in everything. And it delivers an elusive divinity.

CHAPTER TEN

THE CHARACTER OF THE DIVINE

Comprising Diverse Parts

LIKE A DELICIOUS AND RICH STEW, the universe is a whole comprising diverse parts. When you eat the stew, you taste each of the ingredients, and more. The diversity of the universe forms a whole: the stew tastes different from the peas, meat, carrots, and potatoes that form it. The subuniverse unfolds a plethora of bits and properties, which, fundamentally, all relate intimately within the whole. The subuniverse also enfolds the multiplicity it produces, unifying it. This wholeness-with-diversity creates a richness as abundant as you could imagine. As the potatoes, peas, and other ingredients reside in the stew and the stew forms from the mixed ingredients, so the diversity flows from the whole that is the unfolded universe, and the wholeness forms from the mixture. The Divine sponsors wholeness, a unity that promotes diversity.

Thus the separation of parts produces a limiting case, an approximation to the wholeness, a partial knowing of the universe. When considered within the Universe, individual items assume a richness of relationships absent when separate. Within the whole, they discover wealth.

The Divine unfolds this rich universe. Like the stew, the Divine sponsors wholeness. How does the diversity relate to the Divine? How do the individual components relate to the subuniverse whole? The diverse features of the universe in some way reflect the subuniverse Divine that unfolds them. But what does *in some way* mean? How do the attributes of the universe describe those of the Divine?

The chewiness of beef and the sweetness of peas exist within the universe. Similarly the human attribute of love is present. Think of all the properties of the universe and its parts and imagine the Divine as the subuniverse exceeding each of them, exceeding because the Divine unfolds them.

When I refer to the aspects as enfolded, I call them "transcended." *Transcended* applies to all subjective and objective features of the universe and its parts, including all human qualities: our emotions, thoughts, and relationships. They unfold from the Divine as pertaining to the Divine. The subuniverse contains the future, present, and past as well, because it overrides the confines of time and space. Everything and every property, from all times, therefore refer to the Divine.

Using Consciousness as an Example

The spiritually inclined often associate consciousness with the Divine. Many accept consciousness as characterizing humans. Self-consciousness develops consciousness further and, as far as we know, exists only in humans. On the continuum of consciousness, other animals scatter at one end while we sit at the other with our self-consciousness. Here I use *consciousness* to refer to this "self-consciousness" or "self-awareness." Through much of this chapter, this trait is employed as an example of how properties of the universe or its parts fare within the character of the Divine. We can say that, as parts of the universe possess consciousness, so Consciousness— the Divine's transcended version of consciousness—pertains to the Divine. Looking at Consciousness/consciousness as an example, we can ask how the holistic Consciousness of the Divine relates to the consciousness of each individual.

Plato, in a dialogue—more like a lecture—between Socrates and Glaucon, writes:

> "Imagine an underground chamber like a cave, with a long entrance open to the daylight and as wide as the cave. In this chamber are men who have been prisoners there since they were children, their legs and necks being so fastened that they can only look straight ahead of them and cannot turn their heads. Some way off, behind and higher up, a fire is burning, and between the fire and the prisoners and above them runs a road, in front of which a curtain-wall has been built, like the screen at puppet shows between the operators and their audience, above which they show their puppets."

"I see."

"Imagine further that there are men carrying all sorts of gear along behind the curtain-wall, projecting above it and including figures of men and animals made of wood and stone and all sorts of other materials, and that some of these men, as you would expect, are talking and some not."

"An odd picture and an odd sort of prisoner."

"They are drawn from life," I replied. "For, tell me, do you think our prisoners could see anything of themselves or their fellows except the shadows thrown up by the fire on the wall of the cave opposite them?"

"How could they see anything else if they were prevented from moving their heads all their lives?"

"And would they see anything more of the objects carried along the road?"

"Of course not."

"Then if they were able to talk to each other, would they not assume that the shadows they saw were the real things?"

"Inevitably."

"And if the wall of their prison opposite them reflected sound, don't you think that they would suppose, whenever one of the passers-by on the road spoke, that the voice belonged to the shadow passing before them?"

"They would be bound to think so."

"And so in every way they would believe that the shadows of the objects we mentioned were the whole truth."

"Yes, inevitably."[1]

Yes, an odd picture. We perceive, dancing on the walls of Plato's cave, shadows of the real things, traces of the eternal forms that exist in full out of our realm. The Divine exhibits the eternal form of our example, consciousness, as Consciousness. It incarnates as its shadow the consciousness of our experience. For David Bohm, a universal Consciousness occurs in the subuniverse. Each person's consciousness unfolds from and participates in Consciousness, which thereby connects all of our consciousness.

Representing the Indivisible Nature of the Divine

The potential for me to become the eating, sleeping, and thinking human that I am throughout my life existed in me when my mother first felt my stirrings; it still existed in me when my parents presented me with the key to life at my twenty-first birthday; and it continues to exist in me now that my head has turned gray. From outside time, maybe the Divine directed the emergence and development of the universe to produce the properties of it and its parts. Maybe this directing lasts as one long act over the life of the universe. Divine Consciousness could have brought consciousness into existence in the universe, molding what matures here. From the holistic or divine side of Consciousness, we can easily derive the consciousness side of us, the parts.

Before the above vision of transcended properties carries us away, we should absorb this proviso: that the Divine subuniverse constitutes a whole means nothing divides it. All properties of the Divine fuse into one another. We can isolate none of them, either as transcended forms of those of the universe or otherwise. The Divine's character possesses none yet all of the properties of the universe: none because each includes a sense of separateness foreign to the Divine—the Divine is featureless—yet all because each unfolds from and enfolds into the Divine. The Divine is the Divine. We approximate this wholeness with the features we ascribe to the subuniverse, which are in turn the transcended versions of limited attributes. The Divine is the Divine.

The relationships between these three ideas beg explaining:

- the whole that is the Divine,
- transcended versions of properties of the universe and its parts, and
- the properties themselves.

I call upon the word *representations* to help with this. I look out the window at a tree with masses of large green leaves. I could represent this view by projecting it onto two dimensions, as an image on a screen, for instance. This creates one representation. Or I could draw the outline of the leaves on a sheet of paper. This represents the view in a geometrical and frozen way. Or I could paint free-form the different shades of green, representing the tree as a color. Transcended properties similarly represent the indivisible character of the subuniverse Divine, each along a different path.

Relating Diversity to Wholeness

Now we can return to our question: How does the diversity of features of the universe and its parts relate to the wholeness of the subuniverse? We have seen how the wholeness might produce the diversity. But what of the reverse?

For Plato, the eternal forms, the transcended features of the Divine, constitute the real things. For modern Westerners, the universe and its features play this role. For the followers of Plato, the prisoners in the cave wallow in illusion. For most of us, Plato and his followers do. This book's system of spiritual thought asks for a restraining order on both of these approaches. Illusion believes that the experience of the here and now forms the only reality, and illusion believes that reality lies in an order beyond this experience. Reality lies neither in one nor in the other; it lies in both. Both the transcended features and those of the universe are real. Everything that happens and exists unfolds from the subuniverse Divine according to the context there and then. What occurs in the universe represents the Divine in media we can recognize: the colors of trees, the chewiness of food, the caring for children.

All properties of the universe and its parts exist as transcended versions in the Divine. When appropriate, they unfold into the universe to become actualized properties we can experience. All unfoldings from each and every moment of time exist in the Divine, and the way they develop in the universe describes a process that in turn expresses their existence in the Divine. They relate to each other in a holistic fashion whose nature we have yet to understand.

The Divine resembles us. To understand, we impulsively ask about the Divine's character in terms of what we experience, supposing the Divine to display at least some characteristics of humans and other material things. For instance, we understand consciousness to belong to a system such as the human brain and to emerge with the evolution of a neural complex whose parts connect intimately. Consciousness (or a property that produces it incidentally) is adaptive. It enhances the survival of the species against its natural enemies, its rivals, and nature. Matter models mind through evolution. If we approach the character of the Divine from this naturalistic point of view, we automatically seek to understand Consciousness in terms of our adaptive consciousness. Confusion ensues. We stumble around in tangled matters, because unlike the Divine, the evolutionary process involves time. Divine Consciousness can't evolve.

Personalizing the Divine

Familiar terms do not befit Consciousness, though we say it correlates in some way with the evolved faculty called "consciousness." We currently know little about the transcended version of a trait, even in terms of the version exhibited by the universe or its parts. So do we know enough of the Divine's character to say we feel at home with it? The unfolding-enfolding images of the subuniverse and the idea of nonlocality fail to involve us as people. Where, for instance, do they uphold my joys and my sorrows?

These models feel detached and advance our image for the Divine only so far. A black-and-white photocopy of a colored picture falls short of the original because we want the copy to show the brilliance of the flowers in the original. We want to know enough of the Divine to produce a system of spiritual ideas adequate for human experience, a system that involves the whole of each person. Arthur Peacocke thinks that an understanding of the universe coincides with personal meaning because the universe includes us. Our questions of it inquire into our existence and pursuit of meaning. To achieve an adequate interpretation of the universe thus requires personal and human terms for the Divine. The way we understand the universe and our place in it must embrace humanness in order to provide us meaning as its creatures and to involve our full personhood. I hope for a more encompassing symbol. It's *our* divinity.

Suppose we can personalize the Divine. What's the Divine like "in the flesh?" What nouns or verbs describe the Divine? What does a property look, smell, sound, or feel like in its transcended divine form? More generally, how can we know about the Divine?

Think of a living cell in your body. Then try to extrapolate from it to your life as a human being. Too tall an order, you would say. But recall Jurassic Park and the conclusion that geneticists can partially reconstruct the life of a dead creature from its DNA. Perhaps we can likewise approximate the character of the Divine from human and other traits within the universe. Perhaps, with such estimations, we can start to think of the Divine in terms of our world and experience. After all, the world and our experience enfold into the Divine like strands of sacred DNA.

Extrapolating this way projects traits onto the Divine. Yet it's one thing to imagine; it's another for the imaginary to become real. The Greeks could imagine the winged horse Pegasus, but a modern biologist might ask whether a horse could bear a hollow bone structure and the muscles sufficient for the animal to fly without it losing its resemblance to a horse. What would a trait projected onto the Divine look like in its transcended version? How would a characteristic of our world appear if applied to the subuniverse Divine from which it unfolds and to which it enfolds? What's the nature of projections?

We frequently project the human person onto the Divine. This creates a holistic model because humans outstrip the biochemical and physical mechanisms that partly comprise them. The approach of projecting wholes thus suggests three stages. First, we develop models for the properties of parts of the universe. Second, we extend them holistically. And third, we project onto the Divine the whole-part model, including the transcended properties that attract our interest. Appropriate whole-part models found the projections, knowledge for which we pick up from our experience.

We beings and all properties unfold from and enfold into the subuniverse Divine, and so the Divine's wholeness extends our personalness and other traits beyond what we as individuals, as human societies, even as the earth's biosphere can experience. This justifies projecting whatever human or universe traits we wish onto the Divine, for all achieve some accuracy. All ideas apply to the Divine. It's OK to project.

In particular, the Divine is in a sense Personal and Conscious. Several questions and difficulties bear on the way we apply these terms.

• Should we project only human qualities onto the Divine? Many people consider the personal qualities and experiences of humans the highest possible for beings and organisms—or for anything, for that matter. They think we have reached the highest form of existence in the universe. Therefore, they say, we should understand the Divine in our terms; human properties, in or not in their transcended forms, produce the language for the Divine. But current science neither supports nor counters the belief that the human constitutes the "highest" form of existence in the universe. The Divine isn't a person; we just assume its human likeness. The Divine may more resemble a chipmunk or a quartz crystal than a human.

• Meaning and purpose, among other things, link inseparably to biochemicals and genes, to our biology and biological history. Our subjective and personal as well as our objective features evolved. Human qualities, therefore, fail to anchor an adequate understanding of the

Divine, because the Divine didn't evolve; the Divine's character dramatically exceeds that of any human. The transcended versions of our traits look like nothing we would recognize in ourselves.

• Just because the human fetus at a certain stage resembles that of a fish doesn't mean we become fish. To project human qualities onto the Divine doesn't make the Divine much like us. The Divine resembles a person only within the confines of our projections. The Divine isn't so personal, so universally personal as to be a person.

• What does it mean to say the Divine "cares" for the universe, especially human beings? Does Caring (the Divine's transcended version of caring) translate into an activity of the Divine in the universe that we unquestionably notice as caring? If we can't detect it, it's nonsense to say the Divine "cares" for us. Similarly, if the Divine has a Meaning for the universe as we have meanings for our activities, would we clearly notice it? We find it difficult enough to know the meanings intended in works from other cultures and times than our own. La Mouthe cave in the Dordogne, France, contains prehistoric markings, including some of animals and some "enigmatic" lines and symbols, as the French say. But try to read their meaning. That's impossible if we want more than guesses. What, then, of the Divine about whom we know little, not even—unlike the Cro-Magnon artists with whom we at least share a common specieshood—what we possess in common? The chances of our reading divine Meaning are negligibly slim. If we can't recognize the Meaning, it's nonsense to say the Divine has a Meaning for the universe. The Divine creates the universe by unfolding it, but the Divine needn't intend something with this process. A stone does not sense meaning; the Divine may be more stonelike than meaning directed (that is, humanlike).

• We may largely fail to recognize the character of the Divine, despite its including transcended versions of our own qualities. We should cautiously decide what qualities to project onto the Divine.

• Does the Divine sleep and eat? Does the Divine make mistakes? Does the Divine meet moral ambiguity? No. Why not? No basis exists for choosing which personal properties not to apply to the Divine. We can in principle project any traits onto the Divine, and all apply to some extent. The question asks why we should apply some and avoid others. Spiritual thinkers should carefully ponder the range of possible projections and decide why—before they, with us following them, start projecting—we ought to apply this property rather than that one.

In sum, we should project a humanlike character upon the Divine, but we meet limits with this. No quality automatically applies to the Divine. Each needs justifying and to have its limits determined. To know more of the Divine follows a reflective process of trial and error. It requires hard study and solid evidence, and it means trying out our imaginatively created hypotheses against reality. Do they have any truth? Does a suggested trait of the Divine match what we experience both in what we observe of the universe and in our personal experience? Is it transcended and does it fit with the Divine's holism? Does it further our intentions for projecting? We should explore empirically. Also, we need to use care when we project, because projections create social implications. The Nazis and other governments have, for instance, employed certain myths of divinity for their own ends.

While we may, in these and other ways, nudge closer to knowing more of the character of the Divine, the yellow brick road twists on and on over the horizon. Might we relate to the Divine? Once we start to think of the Divine in human terms, we can ask about a relationship between the Divine and ourselves as a species, as communities, and as individuals. If we think we can relate to the Divine, we will want to understand what the relationships might entail, what form these relationships assume. If the Divine shows an interest in individuals, for instance, is this relationship truly personal, as between two people? In relationships, is the Divine a person or is the Divine like a person?

The Divine is like a parent (which means the Divine produces us) and the Divine is like a companion (which means we create along with the Divine). Karl Peters suggests that these relationships between the Divine and individuals emerge from a system of spiritual thought like ours. Peters's observations provide a starting point. But they need drawing out to help us understand how we can engage with the Divine at the deeper dimension we experience with other people. *The Diary of Anne Frank* with its deeply human story and anguish usually engages its readers at a rich, personal level, unlike the theorems of Euclid's geometry and unlike parent and companion images. We need something more than these images in our understanding of the Divine, something that engages living our daily lives with our hopes and hazards, not only our parent and companion experiences.

Relating Personally to the Divine

To pursue the idea of a deeper connection, start with the image of the Divine unfolding and enfolding all our subjective and objective experiences. This continuous producing and embracing means that the root of our personalness participates in the Divine. And this makes our relationship personal. Nonlocality—in particular, the nonlocal correlations between our brains and the Universe—adds to this picture of personal relationships with the Divine. I mentioned earlier that nonlocal correlations appear more in tune with the wholeness of the subuniverse Divine, which we could discuss in terms of the Divine's Consciousness. A relationship of the Divine with a human can become deeper and more personal than a relationship with anything else.

Another approach extends this explanation of personal relationships with the Divine. We understand the world in certain ways, and we project qualities onto what we encounter or imagine. We say the car engine "purrs," for instance. Similarly, we project properties onto the Divine. Because we are persons, we imbue the Divine with metaphors that are personal characteristics. Since the Divine is the subuniverse, the Divine accepts many (perhaps all) characteristics. But the Divine's enormity means that any part or collection of parts falls short of the whole. The Divine doesn't mimic a person, even an ideal person. People see the Divine anthropomorphically, as "the face in the clouds." Each person constructs a divinity out of his or her personhood or the qualities he or she thinks important. This makes the Divine different for each person.

This process of individual projections lays the foundation for our relating with the Divine. We create the Divine's relationship with us. Each person molds the Divine somewhat in her or his own image and relates to that. Forging such metaphors leads us to a relationship with characteristics that appeal to us—the parts of us we see in the Divine—and to which we want to relate. This provides room for different relationships with the Divine and helps us feel connected. In those relationships, we assume a two-way nature with divinity, one that is not just from us to the Divine.

We tend to see the reciprocal side, from the Divine to us, in terms of what happens to us: the Divine "cares" because we escape a flood, recover from cancer, or feel good this morning.

In principle, a person can consciously and actively relate with the Divine in a way that involves her or his personhood. And the Divine does enter such relationships.

Part Three
Searching Morality

CHAPTER ELEVEN

FREEDOM

Identifying the Freedom of the Divine and the Universe

MODERN WESTERNERS THINK HIGHLY OF FREEDOM. We hold it paramount as a virtue for each individual, for each society, and, increasingly, for nature. What constitutes the freedom of the Divine, of the universe, and of human beings? How does the freedom of one impinge on that of another?

A potter produces pots, which then start to do whatever pots do in their lives. They exist independently of the potter. Tradition portrays the autonomy of the Divine and the universe with this metaphor and in this way speaks of freedom. Once fashioned by the Divine, the universe began a life of liberty. It gained an independence to follow its own course, fulfill its own potential, and obey its own laws self-sufficiently. The Divine interferes with none of these.

This belief in the independence of the universe from the Divine led to the theories of early modern scientists, including those of Isaac Newton. The universe exists according to its own laws, free from the willful and fanciful interventions of the Divine. We can, therefore, depend on its consistency. We can experiment with it and achieve the same results each time. We can derive laws that hold in all places and times, because the only other force that could affect it, the action of the Divine, won't disturb it. Science still relies on the result of believing in the universe's freedom.

Yet in our system of spiritual thought, the Divine determines the universe. The universe unfolds from the "doer of all," the Divine. An age-old problem again rears its physiognomy: How can the universe develop and operate, freely following its own laws, when the Divine brings everything about? If the Divine forces the universe to follow sacred dictates, would the Divine arbitrarily interfere in the consistent nature of the universe, like a "cosmic tyrant" gone senile?[1] If so, out would go the laws of the universe.

The universe is free if it can follow its own laws unhindered. This defines the universe's freedom. Does our understanding of the Divine satisfy this condition?

A prong to the definition of the Divine says the Divine acts logically, consistently. The Divine doesn't act willy-nilly. Where I live, we say New England farmers are "men you could set your watch by." Dirt roads, over which many generations of farmers have pulled their wagons and carts, follow the contours of their former fields and created beds that the last century of abandon left unscathed. The universe follows the natural paths of the consistent Divine to create the laws of nature. Freedom roots the universe.

Identifying the Freedom of Humans

Will: "the power of making a reasoned choice or decision or of controlling one's own actions." *Free will:* "freedom of decision or of choice between alternatives" (*Webster's New World Dictionary*). What does our system of spiritual ideas say about free will for us humans? My perspective on it starts, predictably, with evolution: will, including free will, emerges as an evolved property. As such, it's natural. It forms the human face of the freedom of the universe and its parts to follow the laws that apply to them.

Biology influences and directs our minds to believe in our autonomy. Is free will, then, illusory? No; we do possess a degree of freedom. We can choose freely. I chose, for instance, to go back and edit this chapter. But free will partly exaggerates this sense because other factors—biological, cultural, genetic, coincidental—influence and sometimes determine how we act, feel, and think. My choice to edit this chapter, for instance, was in part influenced by my sense of obligation, under pressure from others, to finish this manuscript before deadline. I experience free will, but how to understand it is another matter.

I began to wrestle with these ideas back in my seminary days. The orthodoxy that confronted me there, personified in the seminary's principal, found it hard to resolve the free will–determinism question. Free will appears, according to the principal, when we decide to apply mental energy to a brain process. But that argument failed to resolve my problem: I don't consciously decide to apply mental energy; I just do it. I soon realized, however, that quantum uncertainty offers space for free will. We can't predict the outcome of a quantum state of possibilities. So the supposition suggests that a free choice could pressure the state to become a certain actuality. A free decision somehow coerces one outcome of a quantum event to occur over other possibilities. This happening then causes another event, which causes another, and so on, up through the levels of nature until, finally, the decision becomes a reality. My choice right now to push the M key on my computer unleashed a chain of events: the decision influenced a quantum event in my brain to go a certain way, neurons fired, messages passed through my nerves, and muscles in my arm and fingers moved to push M.

Ian Barbour thinks this approach misses the mark. The uncertainty of the quantum universe bears little or nothing on free decisions. Human free will involves the whole person, and this lies at a level that soars high above quantum events. Too many levels intervene between it and human experience. None of the quantum events that allow my computer to operate, for instance, influences the train of thought you read here. So, Barbour concludes, indeterminism doesn't lead to free will. To think otherwise mixes apples and oranges: quantum events have nothing to do with free decisions.

But, as I previously suggested, nonlocality—a quantum effect—offers a range of options and information to our brain for its decision making. Hence it could explain free will. I also suggested that decisions involve quantum processes in the brain's microtubules via the scheme of Roger Penrose and Stuart Hameroff. Free will includes that creative decision making and so involves nonlocality. It exists in the brain if we construe it as a whole because of nonlocal correlations; this extra property, associated with mind and free will, emerges to exceed what the parts can create. Free will builds from a quantum effect to carry the freedom of the universe and its constituents a step further. This scenario sidesteps quantum indeterminism to call upon nonlocality whose significance lay out of reach when Barbour wrote his critique.

I previously described nonlocal activity as more divinelike than other phenomena, because nonlocality draws more upon the wholeness of the subuniverse than any physical thing we know. Perhaps, with our free will, we especially mirror the Divine. But we reflect the Divine for another reason, too. The Divine unfolds into existence certain of the huge number of possible states of the universe. We do the same, because with our free will, we decide—in a more limited scope than the Divine—which potential outcome of a process unfolds. I decide whether, at the end of my train of thought, this sentence starts with *I* or with *We*. Thus the Divine shares the creator role. When we participate in the activity of the subuniverse and reach into it for an outcome, humans and other beings help create the universe. The Brain/Mind of the Divine includes our brains/minds. Philip Hefner calls humans in this role "created cocreators."

Providing Reality to Choices

The above speaks only about decision making. We lack the power to provide reality to our choices. Deciding to raise my hand and thinking I am raising it doesn't raise it. A choice doesn't achieve anything in the universe; deciding on *I* rather than *We* falls short of typing it out. The Divine, the subuniverse, supplies this power. This is a sneaky point because, like I, you instinctively think that when you do something, *you* actually do it. You think that your internal *I* pushes the mental equivalent to a button, so starting a chain of electrical and mechanical steps that endow reality to your intention. But your or my *I* doesn't do this. I think something. I decide to do something. I think I'm doing something. But I'm not. It's the Divine that furnishes reality to my willing when I act on it. The Divine unfolds everything. It's the Divine that unfolds whatever humans choose to do and then do.

The Divine possesses Free Will (the divine, transcended form of free will), almost by definition. What happens is what the Divine unfolds, including every event in the universe and everything a human wills and does. No power competes with the Divine's. So if the Divine owns a Will, everything that happens agrees with it. The Divine has ultimate free will.

But does the Divine possess Free Will to follow sacred wishes when the universe and we carry on as we like? Is the Divine free to impose the Divine's Will on the universe while the universe can follow its own laws? To add weight to this question, remember that the Divine allows the universe to run with its laws. Worse, the Divine produces the universe running on its laws.

Several points push the confusion further.

• Earlier, we thought the Divine didn't meddle in the universe, that the Divine unfolds it consistently. Does the Divine mindlessly produce the universe in the same way hour after hour, day after day, like a production line for dressmakers' pins? The Divine set up and now operates by the rules of the universe. Do these rules so constrain the divine self that the Divine can't force things to happen a little differently now and then?

• From the Divine everything that happens unfolds, including what we humans do. And what occurs enfolds back into the Divine. What transpires in and to the universe thereby affects the Divine: the pots influence the potter. This creates a vulnerable divinity. Perhaps the limits and dependency of the Divine bind divine Free Will.

• As human free will exists in relation to the determining body (partially determined by genes, environment, and so on), maybe the Divine's Free Will exists in relation to the determining universe. The Divine exercises Free Will within the constraints of the universe, just as we exercise free will within the many factors that constrain us.

• Perhaps, then, the Divine is self-limited. Perhaps the Divine chooses, for whatever reason, to follow the rules. Surely, though, the Divine will overcome all self-limitations to rescue an overloaded ferry from sinking, or a town from massacre, or a child from rape and murder.

To continue this line of reasoning won't resolve this quandary. Rather, we need to revise one of its assumptions without denying that free will applies to us.

The conflict between a free universe and a free divinity that unfolds it only erupts if the Divine can say to heck with the laws and consistency. The confusion requires a resemblance between the Divine and a human being. It assumes that the Divine holds purposes and values that conflict with what we choose to do or with the side of nature that blows ferry-sinking winds. We traditionally and uncritically project onto the Divine the human capacity to value and hold purposes. The disparity between Free Will and freedom alters when we realize this. The Divine enfolds and thus includes human willing, but Free Will transcends free will. Divine wholeness engulfs the transcended version. It becomes unrecognizable. To think of the projection too literally winds our minds up in knots. We understand little of the Divine, too little to know whether the Divine chooses where and how to act.

Further, free will assumes actions in the future, but the future holds no power over the Divine. The idea of free will derives from the flow of time and pertains to the universe we experience rather than to the subuniverse. Thus, in the Divine, neither freedom nor determinism hold sway, because their tension registers only in the unfolded universe.

To inquire about the Divine's Free Will is inappropriate, a categorical mistake similar to asking about the pointedness of flatness. It doesn't exist as we normally understand the term. The wholeness of the Divine produces and respects the freedom of the universe and of each of its parts to be itself but disallows free will for the Divine.

Free will distinguishes personhood, a mark or the epitome of humanness versus machineness. This discussion, then, starts to fill out my spiritual ideas. It begins to touch the nitty-gritty of human life, the experience of being human. It sets off to surpass an abstract image for the Divine that insufficiently relates to our lives.

CHAPTER TWELVE

LIBERATION AND VALUES

Creating and Maintaining Wholeness-with-Diversity

RACIALLY AND CULTURALLY MIXED, U.S. society strives for equal rights for all peoples. Ideally, this would create wholeness-with-diversity. How to do that, however, remains the question. We used to believe in the melting-pot philosophy, but most of us now reject it because those of us from minority cultures want to retain some of our heritage and can't fully eradicate our roots. Nor should we. We have witnessed the failure of integration, court enforced or otherwise. Now segregation raises its head again, this time where minority groups choose segregation as a method to retain and develop in separate ways their equal rights. Will this work? Can a community endure amid subgroups with separate identities but equal rights? No easy or obvious answer exists.

Wholeness-with-diversity emerges as a major feature of our account of the Divine unfolding all. In the diversity, everything fulfills its natural state. This equals freedom. The Divine sponsors the freedom for each to exist as itself, to behave naturally. For humans, our natural way involves making free decisions, exercising free will. The Divine's unfolding roots human free will. But this freedom happens in relationship to all, a holistic freedom where everything connects with everything else in a whole that is the Universe. Such relationships bring out and help create the richness of the diversity and the strength of the wholeness. Theoretically, the United States could duplicate this holistic freedom, at a price: individual and government wishes would collide, requiring lengthy and painful negotiation to resolve.

Freedom-in-connection applies to the environment as well. All things inextricably intertwine with each other and possess the freedom to pursue their lives according to the laws of nature. The wholeness weakens without everything and without each constituent pursuing its own nature. This connection and freedom provide each participant with an intrinsic value, rather than a value born of our use or potential use of them. All possess worth because of their natural relationships, their place in the whole. All beings have the right, says environmental philosopher Ed Grumbine, "to live and blossom and follow their evolutionary desires."

Snapping turtles lay their eggs in sand beside the road a hundred yards from my home. When they hatch, the youngsters forego the difficult haul through the grass to the stream and opt for the easier route across the road. I have yet to see one make it, and crows love the roadkill. In this situation, turtles and we coexist in an unbalanced relationship. Freedom includes our chance to act according to our potential, some of which interferes with that of our partners. We can choose to respect the dynamic wholeness and preserve it—because it is natural reality—or to destroy it. *Ecological wholeness* stands for "wholeness-with-diversity."

In the human world, the idea of the global village is gradually replacing nationalism. The global village represents connections: all of us closely related within a whole. I modify John Donne: no *nation* is an island entire of itself.[1] At the individual level, wholeness-with-diversity implies freedom from political and social oppression. The whole and every part remain incomplete without freedom in a free society; our personal freedom dwindles without the freedom of every person on earth. To render Donne closer to his original, but in inclusive English, "No person is an island entire of herself or himself; every person is a piece of a continent, a part of the main."

Maintaining Liberation-in-Relationship

Holism is also freedom-in-relationship. The whole and every part remain incomplete without each of us living at our fullest, which is living as our inner potential promotes in the context of our social, psychological, and physical needs. It means freedom to be yourself in a marriage that centers on equality and friendship. Freedom encourages the richness of diversity and the strength of togetherness.

Rather than a license for chaos or anarchy where everyone acts as they wish without regard for others, *wholeness-with-diversity* stands for "liberation-in-relationship." Freedom lies within the whole. It requires a balance and an interplay between the desires and needs of the individual and those of all else, including the environment. Freedom is liberation within the nexus of relationships with other people, within the greater environment, within the whole that is the Universe, and within the Divine.

The Divine doesn't force wholeness-with-diversity on us. Freedom and equal rights arise from inside ourselves, and we are free to follow them or to dismiss them. I suggest we follow them, because it's more natural for us to side with our sponsor, the one who stands for freedom. We too should work for wholeness-with-diversity. With our freewill decisions, we can help build the whole the Divine sponsors.

The drive for human liberation builds from the same roots as the drive for liberation for the environment. The causes of the problems are similar, too: the desire for power—power over other people and power over nature. Near-death experiences evidently leave the nearly died with a sense of the wholeness of the world, with a sense of care or love for all people, for all of nature, and a drive to seek wisdom. If that's the case, we should all nearly die.

Employing Power for the Benefit of All

To employ power for the benefit of all poses a problem for most of us. To side with our sponsor and work for wholeness-with-diversity is difficult. The human possesses great potential, Arthur Peacocke writes, but seldom realizes it. Paradoxically, we think of ourselves as the current peak of cosmic development and yet we feel the tragedy of our social and personal shortcomings. Our individual lives are finite but our longings are infinite, Peacocke says. We own a free will. We need to choose which way to go.

Does our system of spiritual ideas tell us which way to go to overcome our social and personal shortcomings? I write and I accept that the Divine sponsors wholeness-with-diversity. I also suggest we side with our divine and natural benefactor to pursue wholeness-with-diversity, that we follow a course where everything fulfills its own potential within the interlocking web that engages us all. "Freedom" and "wholeness-with-diversity" sound well and good as principles, just like "equal rights for all people no matter

their backgrounds" sounds right on the mark. But these ideals run into
problems.

- How do we ensure that turtles reproduce and cars speed to their des-
tinations? How do we make sure that people of different sexual orientations, races, and cultures enjoy equal opportunities? What does
connection mean practically? How do I understand *fullest* when I write
about the incompleteness of the whole and any part without all at
their fullest? How do we achieve these ideals? What do they mean in
life? The wholeness-in-diversity theme needs spelling out in detail. As
it stands, it hardly starts to solve the moral dilemmas of our individual and social lives.

- Nazi Germany was a machine of evil and atrocity that many people
of European descent rightly remember with shame. A together country, as countries go, it enabled some people to play diverse and fulfilling roles toward a unifying goal. Though it failed in the wholeness
vision, it aimed at it. So did Hitler act in accord with what the Divine
sponsors? No. Applying wholeness-with-diversity proves more difficult than we might expect.

This needn't relegate the ideal to the trash, however. We could work
at developing it into practical actions and life orientations.

- We should recognize that we assume wholeness-with-diversity. It
doesn't emerge from somewhere as an "ought" to follow. People may
choose to seek it or not; obviously from the political interests of for-
est loggers and tobacco growers, many people prefer alternate values
to wholeness. Little power stands behind its vision; what would grant
it power? We inquire about motivation, about what would inspire us
to follow an adequate morality.

- We can assume wholeness-with-diversity as a positive value, but ele-
vating it this quickly lacks the support of our system of spiritual ideas.
As yet, it places no value on wholeness or on its opposite, fragmenta-
tion. The subuniverse of wholeness stands the same morally as the
universe of parts, that of our experience. The subuniverse of wholeness
and the universe of parts exist in a constructive tension, and we need
both. In the physics of some chapters back, I considered the classical

approach of parts as an approximation to the wholeness physics of David Bohm and others. This places no value judgment regarding good or bad; physics requires both types of theory. So we ask how the Divine relates to human moral standards.

We need a directed values system, something that applies all through life and leads us somewhere, something that provides a sense of purpose and meaning. To raise wholeness-with-diversity quickly to the supreme good fails our need. It feels right, it sounds superb, it sells well, but the process requires more deliberation, more working out.

CHAPTER THIRTEEN

HISTORIES OF THE UNIVERSE

Laying the Groundwork

THE PREVIOUS CHAPTER ASKED US to search for values to inform our lives in tune with the divinely unfolded universe. I suspect we will find a clue for this in the history of the universe, in the way it has developed since its inception.

Tucked away in reflections on his and David Bohm's metaphysics and physics is Basil Hiley's observation that the beginning of the universe introduced nonlocality. All particles in the universe locked together nonlocally, meaning that all simultaneous quantum events associated with them correlated with one another. Locality had yet to exist. It emerges from the laws of physics applied to an expanding, big bang universe: only when the universe began to expand did fission happen, the particles collide, and eventually locality appear. Locality and separation go hand in hand.

With the expansion appeared the macro universe. Almost everything here relates in a local or classical manner, with the exceptions usually occurring below the macro, at the quantum level. As expansion continues, locality increases. As the universe increases in size, it proceeds from nonlocality to the existence of locality to the dominance of locality. Locality is required for a macro universe and, therefore, us to exist.

The scene Hiley paints depends on the universe not turning back to how it existed previously; the universe proceeds in one direction, plotting an irreversible course through time. "You can't go home again," writes Thomas Wolfe.[1] Drop the concrete block you lifted, and you lose the energy you exerted when raising it.

Locality generally exists at a lower energy level than nonlocality because it represents less organization. To start with, it fails to correlate distant and simultaneous events. Science calls this relationship between energy levels and degrees of organization the second law of thermodynamics: entropy (a measure of disorder) always increases. The universe continuously winds down, scattering its energy right from the initial moment of the big bang.

The terms *locality, nonlocality,* and *entropy* relate in various ways, some of which we've discussed. Other terms require introduction, and further relationships between them merit examination. A picture begins to emerge of two sets of ideas: locality, the macro universe, and separation on one side, and nonlocality and the quantum universe on the other. The histories that await introduction balance the pairing on the nonlocality or wholeness side.

Increasing Complexity

One of the continuing events through time is the increase in separation, which is opposed by increasing complexity, where some parts of the universe join together rather than move apart. This reflects fusion versus fission: fusion of the simple to create the more complex, versus fission and the emergence of locality to domination. The growth of children after they gain some independence from their parents resembles the growth in complexity in the universe. From childhood friendships with a strong dose of self-centeredness, children move out to friends in a self-giving way. Usually, they sooner or later find a life in a close relationship with a significant other. These relationships form with different people and experiences than those of their childhood, and they tie the world of the now-grown child into a web of relationships in which each person retains autonomy.

The universe began with extremely high energy and the tendency to lose it. It started out simple, eventually producing more complex objects, such as suns and planets that store and spend energy. We also see around us biological, social, even chemical and physical systems that increase in energy. A baby starts off small and simple but grows into a large and unbelievably complex teenager.

Ilya Prigogine describes a process by which something can increase in complexity: an unstable system that uses energy and changes chaotically can settle at a stable point with a higher energy level. When you turn off a leaky tap, water at first swirls around the rim of the tap until it gathers enough stability to form a drip. Prigogine even shows the inevitability of such processes, given physical laws.

The growth in complexity of systems such as Prigogine describes assumes the irreversibility in the second law of thermodynamics. In this, it mirrors a requirement Hiley notes for the move from nonlocality to locality. But critics point out that the second law means everything in the universe must run down and, therefore, not grow in complexity. They forget a factor, however. Babies grow larger and more complex by absorbing energy from plants and animals and the people around them. The plants and animals degenerate from structured tissues to the contents of diapers, and parents feel very weary. The complexity of a system increases at the expense of its environment; the surroundings accept more entropy to counter the system's energy growth and stability. So the net entropy of the system plus its environment increases, satisfying the law.

You could think of your body as a collection of organs: brain, heart, kidneys, lungs, skin. These connect with each other to form, with other bits of tissues, the body. But they connect in such a way that the body takes on a life of its own. If one organ suffers, every organ suffers, and the body as a whole suffers. The whole becomes a person who experiences a life impossible for the bundle of organs. That the universe becomes more complex by fusion means some of its parts connect more and more with each other. Different elements join together to form a whole or a system, relating with each other more within the system than when out of it. An increase in a system's internal connections characterizes its increase in complexity. Further, systems may merge together to form supersystems, which may merge to form super-supersystems. Individual people combine to form a community, communities combine to form a region, regions combine to form a country. Systems of systems also demonstrate the increase in the complexity of links.

A whole causes its elements to behave organically in clusters, all together or individually, in ways that differ from how they would act by themselves. In subtle reflections of its wholeness, a complex system directs its parts in its self-regulation, self-maintenance, and defense. These actions of the whole exceed what interactions between the individual parts might achieve. Actions of parts fail to explain behaviors of wholes.

Locality appears when particles start to separate in fission, and later it dominates the macro universe. Nonlocality loses its universality. But does nonlocality emerge at the macro level when the simple fuse to form the complex? Complexity relates separated elements without immediate and physical contact. This could constitute a form of nonlocality, a "macro nonlocality" of the locally related. This nonlocality, though, fails to conform to the instantaneous, quantum type. To avoid confusion, I refrain from calling macro-level wholeness "nonlocal" but instead say a type of connection or correlation or wholeness emerges at the macro level. (To confuse the discussion further, nonlocality at the quantum level may hold together complexity at the macro level. Consciousness occurs to me as an example, as the discussion of Roger Penrose and Stuart Hameroff's model in chapter 7 suggests.)

Wholeness (through complexity) requires the emergence of locality because it needs the parts to link together. Wholeness-with-diversity emerges from separation.

Defining Evolution

Evolution is a history of the universe that builds on both increasing wholeness and the continuing emergence of locality, which creates even greater degrees of wholeness. As a word, *evolution* means several things.

• Usually, it refers to the process called "neo-Darwinian evolution," by which biological species emerge and change. In children's science books, two moths on a tree illustrate the theory of evolution. One moth blends in with the bark and the other stands out. Industrial pollution has dirtied this species of tree, common in the area, and the original moth either changes its camouflage or else becomes fodder for its predators. Natural selection achieves the change in color and pattern by removing the moths with the original coloring and encouraging the variant with the darker markings; those that blend in survive their hunters, those that stand out don't. And the survivors produce offspring with an increased tendency for the darker markings. Neo-Darwinism is a simple procedure that draws together existing processes: a species of reproducing organisms that varies genetically (perhaps by random mutations) plus natural selection, or "survival of the fittest," which works on the variations.

• Neo-Darwinism requires a way for genes to vary. As a source for this, biologists accept random mutations of genes through errors in their production or contact with radioactivity.

• Scientists recently offered several other explanations for mutations, all controversial. Mutations can arise, for instance, spontaneously in cells and bacteria under pressure, as when deprived of nutrition. The organism generates the variations on which selection acts, and, in some cases, this mutating process appears adaptive. Hereditary diseases in humans involve some of these instances.

• Evolution proceeds from an observation or a descriptive theory to become an explanation. We look for evolutionary explanations for phenomena and traits. For example, we explain the appendix as something that evolved to help us digest certain foods at a stage in our evolutionary history but which now serves no purpose. It's harmless, and natural selection has yet to pressure our genes to remove it.

• Biological evolution may require some organisms to move to a higher state of complexity and this may affect the genes, so the change passes on to future generations.

• Hypothetically, suggests Lee Smolin, universes can reproduce through black holes. On the other side of a black hole exists a white hole into a universe and through which energy, stars, and other celestial matter emerge or explode. The more appropriate the conditions for the existence of black holes, the more black holes exist in the universe, and the more universes spin off the parent universe, each with a tendency for the prevalence of the same conditions. Natural selection tends to choose universes with the right conditions for the existence of black holes (or an abundance of black holes) and whose offspring universes possess the black hole–producing conditions. As in this theory of Smolin, neo-Darwinism can apply to other aspects of reality than the biological.

• One type of evolution, not neo-Darwinian, concerns the history of physical things from the moment of the big bang. Cosmology describes such development in the universe, and geology describes the physical development of the earth.

Plate tectonics tells part of this story. The earth's crust divides into adjoining, moving plates that carry the continents embedded in them. About 200 million years ago, a supercontinent named Pangaea covered much of the earth and with plate movements split and resplit into the continents and islands of today. Lines of earthquake and volcanic activity mark the boundaries of plates. One kind of boundary occurs at midocean ridges, where tensions in the earth's mantle pull open rifts, allowing new material to well up and weld to the edges of the plates. When a continent straddles such a rift, it splits apart, forming a new ocean area, such as the Gulf of California. Ocean trenches, a second boundary type, mark subduction zones where plate edges dive into the mantle, which reabsorbs them. A third type of boundary occurs where two plates slide past each other along faults (the San Andreas fault in California, for example). Mountain ranges such as the Himalayas rise where two plates carrying continents collide or, like the Andes, where ocean crust slides under the margin of a continent.

• Oxford zoologist Richard Dawkins introduces the idea of memes, the units of cultural evolution that individuals develop and pass on to others.

> Examples of memes are tunes, ideas, catch-phrases, clothes fashions, ways of making pots or of building arches. Just as genes propagate themselves in the gene pool by leaping from body to body via sperm or eggs, so memes propagate themselves in the meme pool by leaping from brain to brain via a process which, in the broad sense, can be called imitation. If scientists hear, or read about a good idea, they pass it onto their colleagues and students. They mention it in their articles and their lectures. If the idea catches on, it can be said to propagate itself, spreading from brain to brain. As my colleague N. K. Humphrey neatly summed up in an earlier draft . . . , "Memes should be regarded as living structures. . . . When you plant a fertile meme in my mind you literally parasitize my brain, turning it into a vehicle for the meme's propagation. . . . The meme for, say, 'belief in life after death' is actually realized physically, millions of times over, as a structure in nervous systems of individuals the world over."[2]

This sociocultural evolution can progress rapidly compared with biological neo-Darwinism. Changes in genes can take a long time to alter the genetic makeup and hence behavior of all members of a species. But ideas like the computer chip can surface in one day, spread rapidly, and stay forever.

We can chart these evolutionary histories from a human point of view.

- Increasing entropy (the second law of thermodynamics: irreversibility), necessary for:
- Locality, the emergence of the macro universe and therefore necessary for:
- Increasing complexity, that appears necessary for, or at least associated with:
- Evolution in its various forms.

Neo-Darwinian and other forms of evolution result from the expansion of the universe and the wholeness that the subuniverse represents. This is how the subuniverse unfolds through time.

Moving toward Greater Wholeness-within-Diversity

Evolutionary processes seem to move away from wholeness (quantum, nonlocal, whole), but they don't. Two days before I wrote this, news broke of the discovery of life on Mars—incorrectly, it turned out—and a few months ago, astronomers announced the existence of planets outside our solar system. Complexity grows and evolution proceeds throughout the universe. At this stage of its life, the universe moves toward a different type of wholeness than the quantum, though it may still involve nonlocality: a wholeness of internal connections that arises from the evolution of life and other phenomena and from the development of complex systems.

The universe avoids a static status quo that sounds like the conservative politic of "what's good enough for my grandparents is good enough for my grandchildren." The histories of the universe depict the universe moving in various ways, at present toward greater wholeness-with-diversity. A question immediately arises: Moving toward what end point? And that question becomes, What do these histories mean? What's their significance?

Mutations and natural selection represent two natural processes that, when they work together, change the course of the universe. Tensions exist between the laws of the universe and its initial state, and from this tension the universe moves. Change need mean nothing more than this. Change may alter the universe without any larger significance. But why the tension in the first place? Why the potential for "history" versus "maintaining the status quo"? We think meaning must repose there. Another way to phrase this asks, If wholeness remains the aim of and the basic property of the universe, why pass through the process of separation?

Robert Russell asks if the order in Bohm's universe leads to beauty, design, or purpose. If order occurs in what the subuniverse unfolds, does it suggest the intentional design of a creator? Perhaps this design implies that the creator means something by creating the universe. Russell thinks so. I disagree. The subuniverse model inspires a way to talk about the interaction of the Divine with the universe. But I fail to see in Bohm's writings a clear picture of a purpose for the universe, of a movement of the universe toward a goal, or of movement in a specific direction for a purpose. In the Christian approach to history, the universe goes somewhere, from creation to its fulfillment. It moves from its genesis—"In the beginning of creation, when the Divine made heaven and earth"—toward its salvation at the end of time— "Then I saw a new heaven and a new earth" (Gen. 1:1; Rev. 21:1 NEB). The Divine holds a purpose for creation and works through the history of the universe and humanity to bring it about. In our Bohm-influenced scheme, the subuniverse unfolds itself into the universe of our experience and folds the universe back into itself. This flux doesn't work toward a larger purpose. Neither does the movement of a piston in a car engine. Neither does the emergence of wholeness-in-diversity through the processes of complexity building and evolution. The last three words in Russell's question, "beauty, design, purpose," reflect a hope we humans create because we want this as the end point of our inquiry.

Searching for Significance in the Movements of the Universe

We still search for significance in the movements of the universe. Consider the relationship between the Divine and nonlocality. Does the Divine associate more with nonlocality and complexity than with separation and locality? Perhaps connections tune more into the Divine than does separation. As the sponsor of wholeness, the Divine may appear to favor them. But then locality and separation would act against the Divine. We might even call them evil, but, in fact, they don't and aren't. The subuniverse Divine is alocal; the Divine is neither local nor nonlocal. To suggest the evil nature of locality adds an unthought-out moral dimension to the Divine that I want to avoid. While the Divine does sponsor wholeness, including nonlocality, locality helps to achieve it. The Divine promotes locality as much as nonlocality. We need to look again for a basis for value and meaning, because as reiterated at the end of chapter 12, the locality-nonlocality and wholeness–separation pairs fail to reveal divine values in the histories of the universe.

Our abilities, including our mental capacities, evolved. In particular, meaning evolved and is specific to humans. Humans project this property onto the Divine. Events may mean something , but we don't know what this entails. Thus, the histories of the universe possess no Meaning (the divine, transcended form of meaning)—neither does the universe, neither does a strawberry—other than what we provide. To ask for the Meaning of the histories is to project human categories too far at present.

But we need to project. What, then, ought we project?

We evolved into what and who we are. We emerge as a product of the tendency of the universe to increase in wholeness at the macro level, as evolving beings set in the evolving universe. The last chapters spoke of wholeness-with-diversity and the previous chapter said we need more resources on values. Wholeness-with-diversity tells us little about values. Clues to our nature and our relationship to the subuniverse Divine lie in our evolution. Values are social phenomena that appear deep in the life of all humans in community, perhaps throughout the existence of the hominid line. Meaning, purpose, morality, and the content of morals find their roots in our story.

CHAPTER FOURTEEN

SOCIOBIOLOGY

Projecting Human Characteristics onto the Divine

WHAT IF GOD WERE "JUST A SLOB LIKE ONE OF US? . . . What if God was one of us?" Joan Osborne asks in the song "One of Us."[1] People want to understand. So they project human terms and their own characteristics onto a divinity they think exists. The traditional or any other spiritual perspective arises this way. This is why the Divine possesses purposes and meanings. Believers then try to understand what happens in the universe with these projected, now divine, terms. What if God were, Osborne continues, "just a stranger on the bus trying to make his way home?"

Osborne's song emphasizes the humanity of the Divine: traveling home on the bus, acting slobbishly.

> And yeah, yeah, God is great.
> Yeah, yeah, God is good.

The ideals the song perceives of the Divine appear hand in hand with the Divine's humanity. This lowliness we associate with biology. I've noted that biology anchors purpose and the other ways humans find meaning. I also concluded that the system of spiritual thinking I have outlined so far scores weakly on the moral side. Might the roots of morality lie in biology and our evolution?

Accommodating Sociobiology

Sociobiology formed in the sciences to focus the target of evolutionary theory beyond the biological and into the social. It hypothesizes that various genes underlie many, if not all, social behaviors in many species. One way the universe works, develops, changes—one of the histories of the universe—is by neo-Darwinian evolution at all biological and social levels. Sociobiology emerges naturally from this evolutionary perspective. Our system of spiritual ideas thus easily accommodates sociobiology, as a way the subuniverse Divine unfolds. Human sociobiology follows from applying evolution to the specifically human—culture, religion, and so on—as opposed to the generally animal.

I accept and build from Edward Wilson and Charles Lumsden's version of sociobiology, though I remain open to alternative approaches. In some ways, fashion has passed over their version, but the core of their proposal lingers in its competitors. It perseveres as the most well known. Regardless, given what I want to say about sociobiology, it matters little what variation of the theory I describe.

A fundamental premise of the theory is that people succeed, from the standpoint of evolution, when they pass their genes to the next generation. Sociobiology must begin with this reproductive success, also known as genetic fitness. Our interest in sociobiology, however, lies in morality, its origin, function, and content. For a morality to survive, it must improve the reproductive success of those who hold it; it must fulfill an adaptive role for its followers. But how?

Explaining Altruism in Its Many Forms

Parents can promote their children's reproductive success by working hard to earn enough money for adequate nourishment and medical care for them. To improve their genetic fitness, people perform a cooperative behavior called "biological altruism" (what I term "altruism" to distinguish it from regular altruism). "Altruistic" behavior enhances genetic success at risk or cost to one's own chance to produce offspring. The parents I mentioned behave "altruistically" because most of them would prefer to watch TV, munch on chips, and try to pass on their genes. People also practice "reciprocal altruism": While you go to the Red Sox game, I will mind your children to stop them from inflicting permanent brain damage on each other that would limit their chances for producing grandchildren to accompany you some day. I hope you'll do the same for me when I go on a long walk in the woods. Reciprocal altruism happens when people help each other pass on their genes. Someone, sometime, may help them as their reward.

Ironically, then, biological altruism stems from a form of selfishness: the "desire" of genes (to personify them for a moment) to continue into the following generations. This isn't selfish in the normal sense of the word, any more than it's greedy of ants to roll grains of sand up between the bricks on my front walk and irritate me. Ants selflessly do what their genes tell them.

"Altruism" among genetically related individuals sounds reasonable. Reciprocal altruism enables people to behave "altruistically" toward others not immediately related to them but closely connected another way. This too makes sense. But what about "altruism" among peoples who share no blood relationship or cultural or geographical affinity? The existence of this "trans-kin altruism" asks an important question that generates an ongoing research topic for sociobiology.

Goliath's dealings with David show that only some people practice trans-kin altruism. Karl Peters questions its universality and suggests that, for it to exist, only most people need engage in it. This nearly universal practice requires an explanation, providing sociobiology with a way to explore the evolution of morality.

Richard Alexander proposes that morality evolved because it permitted early humans to limit conflicts within groups. They could then form larger communities—which aided them, given their intense competition. William Irons follows Alexander and emphasizes social contracts between competing bands as the origin of morality. With such theories, sociobiology extends its understanding of morality from within a kin ("altruism"), to a circle a little beyond that (reciprocal altruism), to within a larger group and between groups (trans-kin altruism).

Including Epigenetic Rules in the Picture

Sociobiological theory adopts a second hypothesis: The human mind includes various innate patterns or rules by which it works, what Lumsden and Wilson call "epigenetic rules." These rules process information that enters the mind from the outside as well as from internal emotions. They fall into two types. Primary epigenetic rules process raw emotional and sense data. Secondary epigenetic rules assemble inner mental processes, including conscious and deliberate decision making and the placement of values. Epigenetic rules guide people into thoughts and actions that ensure human survival. Genes encode them because they proved so worthwhile in the struggles of our human and prehuman ancestors.

Humans experience altruistic feelings that pressure them to behave "altruistically." These feelings arrive through epigenetic rules and oppose such selfish inclinations as "me first," which also exist for biological reasons. To encourage humans to behave "altruistically," genes guide feelings and moral reasoning. The rules envelop morality with the feeling of objective truth. They can thus enforce "altruism"; humans feel pressure to behave as biology desires. Biology, however, doesn't lock them into blindly following the rules. Altruism is a biologically induced behavior that *encourages* people to behave "altruistically."

The distance from genes to feelings is shorter than you might expect. The stress of driving can cause the release of the hormone adrenaline to stimulate the "flight or fight" response. This helps drivers respond quickly. But it also leaves them snappy and irritable; as they competently weave in and out of lanes, they often demand absolute silence from their children in the back seat. Hormones are chemical messengers that various organs of the body produce and glands release in minute amounts into the bloodstream. Carried to target tissues, they produce either rapid or long-term effects, including alterations of several mental states. Genes connect to behavior by enabling or inhibiting the production or release of such body chemicals.

It is similar for neurotransmitters. A lack of serotonin can propel a person into depression and irritability; the right amount leads to feelings of peace and contentment. Neurotransmitters are chemicals that the bulbous end of nerve cells store and that transmit information across the junction separating one nerve cell from another or from a muscle. When an electrical impulse travels along a nerve cell and reaches the bulbous end, neurotransmitters release and travel across the junction, either prompting or inhibiting continued impulses. Genes regulate these processes, too. By inclining mental attitudes, genes suggest to a person what to do.

Evidence for sociobiology continues to mount. Papers and newsmagazines, for instance, regularly publish items that suggest it.

• A gene normally initiates the flow of dopamine, a brain neurotransmitter that provides people with their sense of well-being. But if defective, the gene can diminish the dopamine function. This may drive a person to drugs, alcohol, or activities, such as murder and other violence, that boost the dopamine affect.

• A small number of people with brain injuries experience such behavioral problems as inappropriate actions with the opposite sex, interrupting conversations constantly, stealing, or hitting. This happens because the part of the brain that controls inhibition and impulse—and distinguishes between wrong and right—lies just behind the forehead and is thus frequently damaged in head injuries.

• The work on split brains by Dartmouth neuroscientist Michael Gazzaniga shows the left brain to house the main elements of our ability to interpret our own and others' behavior and emotional states and to infer how the world operates. A left-brain mechanism seems to exist that constantly seeks relationships and that assesses where we stand with others. Gazzaniga dubs this the "interpreter." He has also found that the left brain helps feelings to break into consciousness from trains of thought automatically running through our minds. It can influence memory, sometimes for the worse. And it helps considerably in problem solving.

• Chimps share food and care for others in their troop on the understanding that the others will later return the favor. The troop punishes those who fail to meet this obligation. This reciprocity and obligation closely approach what we call "rights," with the idea of fairness close behind. Chimps possess the emotional framework for morality.

• Twin studies in process by Lyndon Eaves and his colleagues at the University of Virginia show the high correlation of the social attitudes (including spiritual inclinations) of identical twins, whether reared together or apart from birth, in comparison with nonidentical twins, siblings, parents and children, and unrelated people. Genes appear to influence certain aspects of culture, including beliefs and attitudes.

• Now researchers can identify an individual gene that helps shape excitability, a specific trait of the personality.

Overcoming the Selfish Tendencies Produced by Genes

Biology plays a major role in behaviors we normally associate with culture and individual choice. What part might *culture* then play in behavior?

Donald Campbell thinks nature tends not to select "altruistic" tendencies that put an individual at risk. It prefers those behaviors from which the individual selfishly gains. This limits the genetic evolution of biological altruism and hence the development of altruism from biology. Instead, Campbell suggests sociocultural evolution caused the altruism essential for contemporary urban civilization. Only it could overcome the selfish tendencies the genes produce. Religions, he adds, play a major role in this evolution.

Ralph Burhoe's approach to sociobiology similarly says that "altruism" can explain only so much. Culture types—socially inherited packets of information (including religion and language)—shape societies. You could think of them as nonbiological cultural genes. Richard Dawkins calls them "memes" and cites the idea of God as an example; each generation for many thousands of years has passed to its successor its understanding of who God is, sometimes slightly modified from experience and insight. The environment selects the memes independent of, but in harmony with, biological genes. Religion remembers and culturally passes on to the society long-range values and goals, and natural selection works on the religion to remove harmful elements. In this way, religious goals and genetic goals depend on each other. Religion creates the altruism necessary for a society's development and sustenance.

Biology starts culture off, but culture assumes control over its direction through sociocultural evolution, building on the biological to help create what we are today socially and as individuals. Campbell and Burhoe, therefore, stress an evolution other than the biological one as the way to understand the development of culture. Their ideas may contribute to knowledge and add to what we can know with sociobiology. If so, their theory of sociocultural evolution should tie in more closely with the research of such people as Alexander who explain trans-kin altruism biologically.

It makes more sense to say genes and culture work together. Wilson and Lumsden call their scheme "gene-culture co-evolution," where

- genes affect the direction of cultural change;
- natural selection winnows through the change;
- culture exerts some of the selection pressure;
- natural selection shifts the frequencies of culture-informing genes;
- new channels for cultural evolution then open up; and
- the circle carries on around again.

Biology and culture are inseparable. They evolve together.

Culture constructs a society's moral system, based on biological requirements. The culture determines good and bad by comparing various epigenetic rules and then builds, sorts, and develops the genetic impulses or epigenetic rules into a coherent morality. Sometimes culture even shapes the resultant morals to oppose their original biological function. Sociobiology believes that culture and the individual can and do change and add to biological norms; it doesn't say genes control all behavior in a mechanical fashion. Rather, genes inform culture flexibly. Ethical codes can differ, and sometimes they oppose what biology would require—monogamy can oppose procreation, for example.

Assessing the Roles of Biology and Culture

To what extent does biology play a role in morality, and to what extent does culture do so? Wilson writes, "Can the cultural evolution of higher ethical values gain a direction and momentum of its own and completely replace genetic evolution? I think not. The genes hold culture on a leash."[2] While the leash is long, it does constrain values. Culture exceeds biology, but biology always tethers it.

Others like Arthur Peacocke disagree with such a large role for biology. Genes don't determine most social behavior, he thinks. While he admits that research may confirm sociobiology, he rejects the possibility that genetics will explain all of culture. Sociobiology poses no problem for belief in the Divine, says Peacocke, if it only partly explains the person. Such a theory could even describe a way in which the Divine creates. But, for Peacocke, culture involves unique, emergent questions and behaviors. The leash stretches a very long way. So long that we might as well say culture breaks away from biology and wanders free.

So how far does the leash extend? To what degree does culture build from biology? Robert Plomin reports that while genetic inheritance plays a major role in behavior, the genes of many species account for at most 50 percent of the variation in their behaviors. Perhaps this 50 percent holds for humans, too. What proportion of our behaviors arise from our genes, how much from choice, how much from an individual's personal and social idiosyncrasies, and how much from culture?

This question suggests we can assign a quantity to each, that we can draw a line between biology and culture. Some writers think biology pervades below some level and culture above it. But no clear-cut dividing line exists. Despite their nonbiological appearance, people's feelings around cultural attitudes and their reasoning can result from biological pressures, epigenetic rules. John Bowker thinks it "a very distant goal" to say more precisely how much genes determine and how much culture constrains. The hopelessness in his view arises from the complexity of the relationship between cultural and genetic evolution. I more than agree. Rather than say it is "a very distant goal," we can more precisely say we will never, in principle, attain that precision. The genetic and the cultural will remain inseparable.

Unknown to him, Peacocke's stand over objectivism and subjectivism provides a model that solves the biology-culture problem. Do the findings of science accurately represent the universe independent of us, or are they mostly human creations that reflect back to us the ways we gather knowledge? Peacocke calls his solution "critical or qualified realism": scientific theories portray the structure of reality, but sociological and psychological factors play a role in them. His punch line adds that we can't isolate, let alone remove, the subjective effects from the theories. The objective and the subjective mingle so much that they become an inseparable unity. This epistemological solution by Peacocke reframes the whole biology-culture boundary question when we replace objectivism with biology, and subjectivism with culture. Genes determine human behavior, but cultural factors play a role as well, and we can't isolate the cultural effects from those of the genes. Biology and culture work so closely together that we can't separate them.

Biology and culture so infuse one another that what we examine is a whole, biology-culture. Given my usual approach to the mutual relevance of scientific and spiritual thought, you can guess that I want to promote a close and constructive relationship between sociobiology and spiritual thought, between genes and culture.

Culture acts as an instrument of genetic survival. Sociobiology says biology starts and drives any cultural activity, partly directing it with epigenetic rules. Meaning intertwines with the biological; genes encourage spiritual and other cultural activities to discern and promote what we aspire to, to raise and answer the questions of meaning . Biology may even lead people to believe in the primacy of meaning. For Peacocke, the idea of the Divine belongs to the larger framework of meaning. It grows from questions such as the following he lists: "Why is there anything at all? What kind of universe must it be if insentient matter can evolve naturally into self-conscious, thinking persons? What is the meaning of personal life in such a cosmos?"[3] Human biology inclines us to assemble a larger framework of meaning than the responses of science might provide, and it may influence the content of our framework. Culture adds to what the genes bring, helps enforce what biology requires, and in the long run, helps choose the genes through natural selection. Biology makes us make meanings for our lives.

Making a Case for Sociobiology

Our biology wants us to exercise choice in that meaning. Some who want to perpetuate existing injustices claim the intellectual inferiority of one race to another. This status, they say, arises from evolved human nature; evolution produced it and we ought to accept it. Sociobiology evoked controversy at its inception in the 1960s and continues to—many spiritual thinkers still oppose it in principle—because critics sense the dangers of its misuse. They accuse sociobiologists of promoting those dangerous and dehumanizing views.

Both the racists and their critics warp sociobiology. Sociobiology does not support this racism as the only or the chief interpretation. Rather, it permits the opposition of injustices. Biology prompts what most of us would judge as both altruistic and evil inclinations, and it falls to social reflection to discriminate between them and to emphasize the more appropriate (perhaps the altruistic, the anti-injustice) behaviors.

Researchers now frequently shun the term *sociobiology* and opt for other less controversial terms—*human behavioral ecology* and *evolutionary psychology,* for instance—which avoid reductionist tendencies and overreaching claims. I decline this trend here. How we employ the theory provides the problem—and the opportunity.

The Human Genome Project nears its completion and increasingly draws causal links between traits—physical, emotional, thinking, behavioral—and genes. The potential for deterministic applications and abuses of the results of the research challenge us. They raise moral issues. For example, what will we do with unborns who show a defect in the gene that influences the flow of dopamine, if we want a society free of murder and violence? Abort them? Social-biological engineering shouldn't perpetuate or initiate social injustices. When geneticists fully tag the human genome, we need a large and solid body of spiritual reflection to inform the application of this information. Approaching human sociobiology positively and seriously will help.

A window of opportunity opens for spiritual thinkers to explore this phenomenon, to let down their defenses and work constructively with it. They could agree that morality and other human properties depend on biology. And they could waive the need for sociobiology's complete veracity before they engage in this task. I feel anguished when I read such delaying tactics in scholars' reflections; sociobiology enjoys sufficient support to offer more than idle speculation, and continuing research affirms it. Even the well-confirmed and full science of classical physics suffers limitations and ambiguities. Sociobiology won't disappear. In whatever form it materializes in the future—for it will mature to be different from how it is presented here—it will offer many resources to spiritual thought on such issues as morality. We can approach it now with seriousness and openness, aware that some of our current conclusions may change.

CHAPTER FIFTEEN

DOES MORALITY
COME FROM THE DIVINE?

Searching for the Source of Morality

To STRENGTHEN THE MORAL ASPECTS of my system of spiritual ideas, I introduced the discipline of sociobiology in the previous chapter. Biology forces us to seek out and find meaning for our lives and the world we live in, and part of this meaning involves morality. Biology requires morality. Through epigenetic rules, biology also helps provide the feeling of rightness to these virtues. Does it also supply the content of the morality?

Or does the Divine supply this morality? Does the Divine follow a system of principles—or think one up—and reveal it to us? Perhaps the Divine does and we have yet to hear it fully. Perhaps we should search it out. I noted that the wholeness-with-diversity that the Divine sponsors fails as a morality; but should we look more thoroughly for details or listen more keenly to its originator?

Questioning the Divine Morality

This discussion on sociobiology and morality centers on altruism and biological altruism. Biological altruism emerges as a key notion for the traffic between spiritual ideas and sociobiology. Enter Canadian philosopher Michael Ruse with his hard-hitting comments on sociobiology and spiritual beliefs.

First, Ruse turns to what he calls the "love commandment": "Love your neighbor as yourself." "Real Christian practice," he believes, centers on this command. He distinguishes two interpretations of it, a weak and a strong form. The latter requires us "to love everyone: family, friend, nodding acquaintance, and enemy." Further, we must forgive our enemies "virtually without limit."[1]

Biological theory and empirical research suggest to Ruse that "altruism" exists between our kin and us (we want our relatives to reproduce). It also exists toward people with whom we could exchange help (reciprocal altruism). However, the more distant others are from our immediate circle, the less we feel morally responsible for them. This conflicts with the strong love command and undermines it as an ideal toward which to strive. While the love command and biology agree on some level of altruism, they disagree on extreme forms.

Even worse, Ruse continues, the biological urge to retaliate undermines turning the other cheek without limit. Biology encourages frustration at abuse. Humans by nature seek to counter maltreatment near its onset. Thus, the strong love command acts against survival.

In the previous chapter, I noted that the idea of epigenetic rules helps explain why people hold to moral ideals despite other biological impulses. Richard Alexander's development of sociobiology tries to explain trans-kin altruism. Ruse doesn't deny such theories, but points to the difficulties of the strong love command as a value toward which we might aim. A moral dictate must by itself appeal or make sense to people for them to accept it, and biological morality corresponds more closely to usual practices and instincts than does the strong love command. Thus Ruse finds good reasons to adopt his evolutionary position rather than the Christian one. He points out that most would think it irresponsible to allow someone else to sin against them 490 times, as the Gospel of Matthew mandates. He feels uncomfortable about a divinity who requires morally strange behavior.

Ruse can't accept the strong love command as necessarily involved in altruism or biological altruism. His application of sociobiology starts to question divine morality because Christians assume that the strong love command emanates from the Divine, lies at the heart of the Divine's own morality, and centers a way of life we should follow. Biology, however, affects the content of the larger framework of meaning that includes morality.

Ruse then bypasses all love commands and, adding to his case against Christian belief, opens up the heart of his objection. He believes he can undercut all forms of Christian altruism or any other kind of altruism. His argument applies to any form of spiritual belief where a divinity holds moral power over adherents, and it cuts into the base of ethical claims in three movements, the first two of which I have just described.

- Morality is a biological adaptation. We perceive morality, sensing right and wrong.

- For morality to work, it must feel natural and claim power over us; we must know it defines the right and proper way to live. Through epigenetic rules, we feel this obligation to behave rightly.

- I want to do the best for my kids. But if I'm starving, I'll also want to eat my fill first. Some epigenetic rules endow morality with a naturalness and power, but others provide clout to feelings that conflict with moral positives and could undermine them. Ruse raises this point when he shows the difficulty in holding the strong love command of Christians.

For a morality to work, it needs an extra kick other motivations lack. The additional power arises by a natural process in which believers project the morality onto a divinity (the All Powerful). They, then, believe in an independent and objective moral code that is changeless and independent of human conditions. It emanates from something higher than and outside of themselves. Feeling this absolute, moral other as a force on them, they follow its moral dictates. They believe their divinity requires it of them and they strive to obey it.

Ruse thinks biological explanations of the processes involved destroy the Christian belief in the Divine's morality. To recognize morality as only a biological adaptation undermines its traditional support. As Shakespeare writes, "Ay, there's the rub."

To raise his challenge, Ruse enlists his reading of biology-based sociobiology. We have seen that culture plays a part in deciding what believers project onto the Divine and what they project onto the Divine's dark shadow, what we might call the "devil." Culture similarly plays an essential role in transforming biological altruism into altruistic behavior. Does Ruse's challenge fade or disappear when we accept a role for culture? Several scholars—including Donald Campbell and Ralph Burhoe, mentioned in the previous chapter—think altruism so transcends biological altruism

as to leave its gene-held leash. They stress sociocultural evolution over biological evolution. Burhoe says a spiritual tradition functions to remember
and culturally pass on the society's long-range values and goals. It, and not
biology, promotes the altruism essential for the development and sustaining of society.

Ensuring Survival through Morality

Altruism concerns survival, both a society's survival and an individual's
genetic survival. A spiritual tradition portrays a divinity as the source of
altruism and thereby instills altruism in the members of its society. The
Divine functions to achieve this. Morality refers to nothing more than this.
Though it stresses sociocultural evolution over biological evolution, Burhoe's
theory still allows Ruse's challenge. Both sociocultural theory and sociobiology produce the same outcome: they unmask the morality of a culture's
divinity as a projection to help its members' genes survive. Ruse provoca-
tively adds, "Morality is just an aid to survival and reproduction, and has no
being beyond or without this."[2]

The above describes how morality relates to a divinity, including the
Divine that our system of spiritual ideas defines. It doesn't prove the
Divine's or any other divinity's nonexistence—as if the Divine is nothing
but a projection of moral and other sentiments—despite Ruse's atheistic
intentions. What does it tell us about the Divine, then? While we project
moralities onto the Divine, perhaps the Divine actually possesses one. Suppose, then, that the Divine does adhere to something we would recognize
as a system of moral values.

> • The Divine creates forms of life through evolution. Perhaps the
> Divine worked with our ancestors' genes to implant the divine moral-
> ity in them. Or perhaps we evolved into the Divine's morality. Yet the
> biological foundation of the moralities to which humans aspire is
> unique to humans. Human morality arose in part from purely chance
> events (random genetic mutations, for instance) and from survival
> pressures peculiar to our species. These pressures placed no constraint
> on the Divine who also exerted little or no control over the chance
> events. Our morality lies in a different league than the Divine's, if the
> Divine possesses one.

• The Divine's morality appears as a tendency throughout all existence because, the universe unfolds from the Divine, moment by moment. The Divine's values would infuse into all species. Yet cannibalism (the praying mantis) and incest among many species, for instance, differ from human ideals. And some people find or found cannibalism acceptable. Different species and different groups within our species hold different moralities. No virtue stands out as universal. Thus the Divine bears nothing we would recognize as a system of moral values.

We project moralities onto the Divine. The Divine enfolds the moralities humans and other species hold (assuming we can apply this term to non-humans) and, as with other personal or subjective properties, the Divine holistically surpasses these properties. The Divine may behave according to some standard (can we even apply the word *standard*?) that to us looks nothing like a morality.

I press this further. The Divine doesn't align with any morality, whether we would recognize it as a set of values or not, because morality is an adaptive feature. The Divine doesn't include a digestive tract because we do. The Divine is amoral. We must, therefore, look elsewhere than to the Divine for a morality to emulate. Where might we look?

CHAPTER SIXTEEN

DOES MORALITY COME FROM BIOLOGY?

Striving to Survive

BECAUSE THE DIVINE DECLINES TO DELIVER OUR MORALITY, where might we look for it? Even though it belongs to the mind, perhaps the content of morality relates to something physical happening inside us. Perhaps biology offers the best place to look, given the biological rootedness of this human phenomenon.

The emergent features of social behavior appear from basic requirements set up long ago by evolution. From biology, people can interpret themselves to themselves at their own state of cultural development. I understand my psychological condition in terms of the biochemicals and electrical impulses that operate in my brain and body rather than in terms of a failure or success of my noumenal will or the state of my eternal soul. Biology grounds mental aspirations by requiring a framework of meaning whose emphases lead to biological altruism, by providing the tools to create it—reason and reflection, for example—and by influencing its content. Edward Wilson writes:

> The desire to preserve our own lives or the lives of those to whom we are closely bonded by love—our families and tribes—is the deepest trait one can find in human beings and animals. . . . If there is any single value that is fundamental to all life, it is the struggle to stay alive as a species.[1]

Survival as a species produces our strongest drive. To survive, we must create meaning, and so survival becomes the main value or meaning toward which we strive. (Perhaps it centers a spiritual outlook.) Within our frameworks of meaning, each of us works out the significance of our biologically rooted values. This world of meaning may nudge us to ask, in the words of Arthur Peacocke, "Survival for what?"[2] Some Christians, Peacocke's argument continues, believe that humanity primarily serves to glorify God and that God may intend something other than our continuation. This framework of meaning (biology promoted, as are all frameworks of meaning, according to my discussion) denies that our survival as a species is the most urgent value. A meaning system that says this has gone haywire; it has lost its consistency. The question, "Should I switch on the cruise control?" only makes sense when I drive the car. The question, "Should we survive?" only makes sense if we survive; "survival for our nonsurvival," to paraphrase Peacocke, is surreal. In the end, questions of meaning fall secondary to survival and sink without it.

Our biology promotes our genetic survival and the world of meaning in which to discuss and push for it. So I disagree with Thomas King when he writes, "Science has provided us with much, but it will give us an ethic on the same day that it gives us a square circle."[3] I also disagree that a spiritual approach can only add "emotional fuzz to values developed elsewhere"—as King thinks Wilson says.[4] Our genes provide us with a rich context in which to live our lives: they require and inform the raw animal in us, and they require and inform the qualitative or aesthetic in spiritual life, in thought, and in experience. Our spiritual and moral lives exult, in the words of John Bowker, as "tunes sung by the genes."[5] These all serve, as their basic function, the survival of the human species in its fullest, the richest orchestration possible.

Informing the Content of Morality

At least since David Hume, many have argued for or assumed that an infinitely deep and wide gulf separates what humans are naturally and how humans ought to behave—which is opposite to the approach I pursue. The use of sociobiological knowledge to help inform the content of morality is, therefore, radical and requires further support. Of the many reasons buttressing it, the following stand out especially:

• Several years ago, my health urged me to think more seriously about what I eat. From inquiries and from what I read, I ended up with the diet that, it turns out, nutritional science recommends: we should eat foods at the base of the food pyramid more frequently than those at the top. Breads, cereals, rice, and pasta sit at this base. Above these belong fruit and vegetables. Next come dairy products and a group including meats, eggs, nuts, and dry beans. Oils and sweets ascend to the apex; these we should eat sparingly. Though difficult to follow when I travel, I feel better on this diet and, my doctor says from blood tests, it leads to a sounder me. Developing a morality with the assistance of sociobiology equates to developing a diet with the assistance of nutritional science.

• Scientific investigation (say, sociobiology) explains how humans behave. Call this the "is." People used to believe in the absolute truth of scientific knowledge, and many still want to safeguard the more subtle and vulnerable knowledge of the spiritual and moral from aggressive science. The "oughts" of morality require immunity from the "is's" of science. Protectionists fear the reduction of culture to the results of biological mechanisms. But the threat dissipates with the disclosure of this "scientism" as a sham. "Oughts" can stand beside their equal partners, the "is's."

• I wake up early in the morning, eat breakfast, and start work whether I feel sleepy or not. This says something about my dedication and regularity. It says something about my relationship with my writing and the values that drive me. The Divine unfolds humanity along the evolutionary path, which means sociobiology describes one of the Divine's ways of working. The "is" must, therefore, say something about the Divine's relationship to humanity. To fill out the "ought" from a spiritual point of view, therefore, will draw extensively on the "is."

• Some scholars conclude that the "is" plays no role in determining the "ought." Genes exert no control over culture, nor do they contribute to it. We should fix a gulf between the "is's" and the "oughts." They lie on different shores of the ocean of life and will remain apart forever. The scholars think this because they believe humans decide their own actions with moral reasoning. By distancing ourselves from our inclinations and instincts, they say, we can impartially work out our motives and reasons for what we do.

Detached scientists in white coats examine but never disturb separate and objective reality: many of us learned this myth during our school days. But as we've seen in the last decades, this belief about science lacks support from its history and the way scientists work. Thus I emphasize the commonalities of the "is's" and "oughts," believe in migration from one to the other, and anticipate their wholeness when, ideally, they work together. We should apply "is's" to inform the "oughts" where appropriate. But then we must remember that scientific knowledge may change. Any morality suggested by contemporary sociobiology could and may well appear inappropriate to future generations.

• Teenagers follow their natural biological path when they behave outlandishly and question their parents' authority. They think they express their individuality and originality, but they follow a powerful norm. Though they think otherwise, ethicists similarly obey the "is" when they debate what the "is" means, when they try to obtain "oughts" from "is's." They also follow it when they discuss what an "ought" implies for different situations. The "is" requires deciding on and amplifying the "ought." Further, through its epigenetic rules, the "is" partly directs what the "ought" means. Genetic survival requires and informs moral decisions.

• Some cultures embrace totems, usually plants or animals, one of which each social group or individual reveres. A group totem represents the group's bond of unity, perhaps the ancestor or sibling of the group's members. The pelican stood for the totem of the Narrinyeri Aboriginal in A. P. Elkin's story in chapter 7. Whoever the peoples and whatever their totems, incest is the prohibited marriage between those of the same totem or the same clan.

Incest taboos crop up throughout cultures, according to sociobiological literature. Common moral elements appear from one people to another. Sometimes an "is" can become an "ought" and probably should do so for a morality to function as biologically intended. In particular, biological altruism and biological reciprocal altruism form the base for each society's morality. Cultures will amplify these norms, provide them with various degrees of importance, and express and qualify them differently. Biological altruism and biological reciprocal altruism nonetheless define the good, the "ought," for all peoples past and present. They should found the good for future peoples as well.

No strict division exists between "is" and "ought." The "is" informs, but never dictates, the "ought." Within limits, we can and should use sociobiology to help in moral decision making.

Some of us can as easily become angry and lash out at sick or infirm friends as feel compassion and try to help. Evolution provides us with conflicting motivations. That is the "is." In many situations, the "is" fails to say precisely how we should behave (the "ought"). Relying on natural motivations fails to carry us all the way; not all natural things are good. Thus creation leaves us with decisions as to which of these inclinations we should follow. For this, we employ our inborn faculties: not only our feelings and the effects of epigenetic rules but also, for instance, our skills of reason and judgment. Is it better to express anger at a sick person or to console her or him? What if the sick person is your ten-year-old who has overdosed on soda and chocolate? The process of meaning-making or morality-defining also requires input from sociobiological knowledge to say what is necessary for a morality, for the morality—for instance, codes that provide biological altruism. Scientists, ethicists, and moral reasoners help decide how we should behave, given the competing options open to us.

Reacting to the Biological Rootedness of Morality

Many spiritual thinkers react with fear to the biological rootedness of morality. Who's to say what's in the morality and what's not? Perhaps anything goes. They'd prefer that ethicists informed by spiritual beliefs produce uncontaminated decisions. Michael Ruse and Wilson describe morality as "an illusion fobbed off on us by our genes to get us to cooperate. . . . It's a shared illusion of the human race."[6] Their provocation fuels the fear.

King applies Ruse and Wilson's argument to sight. The eyes help us survive, but what they see isn't, therefore, an illusion or nonexistent. You could equally apply the argument to a sociobiological analysis of science. Science too serves biological functions enabling survival. What it says about the universe isn't, therefore, an illusion or error. That morality helps us survive doesn't make it an illusion.

The word *illusion* provokes the word *arbitrary:* nothing independent exists against which to judge a morality. What if you can change the word *illusion* to, say, *means:* "Morality is a means used by our genes to get us to cooperate. . . . It's a means shared by the human race." Changing the word still leaves the question of arbitrariness, however; it only helps dampen the inferno Ruse and Wilson fan. Is there a way to counter the sense of arbitrariness when it comes to morality, given this sociobiological perspective? We revisit the lesson of chapter 14 over how much of culture genes determine and how much humans invent.

Look at science again. Our humanly contrived models constitute scientific knowledge. So how science understands the universe is peculiar to humans. Horses, for example, wouldn't exhibit or need the slightest interest in molecular physics. Science exemplifies the human capacity to know and it helps us understand and thereby control our environment so that we can more ably survive within it. But the universe still exists. Further, our models in some ways resemble it.

Or look at our eyesight again. It functions to provide sensory input from our physical environment so that we can better perform and thereby survive within it. What we see is peculiar to us. What we see relates closely to what exists out there, but our human apparatus conditions it. I'm red-green color blind. The red apples on the tree exist to me only when I walk to within a yard of its branches. What I see relates closely to what exists out there, but my peculiar human apparatus conditions it. What we see inseparably intertwines with how we see it.

Now look at values. The biological rootedness of morality means that values throb away inside us. The ones you possess may enable the survival of your genes, and, in their roots, they transcend both you as an individual and us as a culture. They encourage you and me to relate to each other and our environment so that our race may better survive. This defines the sense in which a morality isn't arbitrary and against which we can judge its truth: it must enable what human continuation requires, in the full sense of *human*. A morality must advance the behaviors that will best enable the survival of humanity, a survival that embraces the richness of the cultural and meaning frameworks that pump at the heart of who we are as a species.

Suppose our culture worked out a morality. It might appear we forged it after our own image, that we molded a golden calf. But that wouldn't be the case. Our biology stands prophetic; it rears up against us and tells us when we err and when we stray. The prophetic force thunders from beyond us and our morality, from deep within our existence. It owns moral otherness.

The ideas of sociobiology interact at face value with their spiritual counterparts. Sociobiology offers a lot to spiritual thought. And, as an aspect of evolutionary knowledge, it ties in well with our spiritual foundation. It could help fill the hole in the approach of this book by providing a foundation for morality. In fact, one could say there's no moral hole in the system of spiritual ideas this book develops; we only need to continue developing the ideas.

CREATING A MORALITY

Lacking an Imperative to Behave Altruistically

BIOLOGY REQUIRES US TO FORM AND HOLD TO A MORALITY, and it also informs the content of that morality. Though the last chapter discussed that topic, one remaining factor about a morality still begs a closer look: the strength of convictions. What forces so many of us to hold tightly to moral beliefs such as "to have and to hold, from this day forth"?

Some years back, the front-page headlines of a newspaper read: "Deadly Air Bags. Key Finding: Two Children Could Be Killed for Every One Saved." "Delta Engine Hub Was Split in Two Places." "8,500 New HIV Cases Occur Daily." "'Epidemic of Violence on the Job' at All-Time High."[1] Many forces threaten human survival. We seem unable to act and solve many of our problems and we—at least some of us—lack the moral will to do so. Just think of the wars in the former Yugoslavia and Chechnya. We often lack the morality that says such and such needs doing as well as the moral courage to do it.

At least two sources drain the knowledge of what altruism means in daily life and subvert the imperative feeling to behave altruistically.

• Altruism in the West used to arise in part from obeying the strong love command, "to love everyone: family, friend, nodding acquaintance, and enemy." With the growth of secularism and the consequent demise of the power of Christianity and Judaism and their divinities, down spiraled the support for the command and thus for altruism. We now face relativism, recognizing many of the different claims and interests that clamor for service. The diminished power of traditional spiritual institutions allows many of us to do just about anything, even interfere with institutions previously thought sacred—marriage bonds, for instance.

• Michael Ruse's case against Christian altruism helps us recognize the biological vehicle behind morality. We create it under pressure from our biology, project it onto the Divine, and then read it off as a requirement for us. Just as naming an ailment and knowing how it works helps me handle it, this information about altruism undermines Christian altruism because now we can question the power behind it.

The issue goes further. Mothers who use drugs or alcohol tend to give birth to offspring with physical and mental problems who are, when their turn comes, less likely to reproduce successfully. If it exists, the force of altruism can help the mothers stop themselves from becoming or staying hooked. Biological altruism depends on the force of altruism. Sometimes biological altruism needs a push from powerful altruistic feelings to counter other motivations such as feeling good on drugs. Undermining the altruistic feelings, therefore, also undermines biological altruism. To undermine biological altruism, suggests sociobiology, threatens our survival.

Seeking a Modern Content for Altruism

Despite secularization and relativism, humans still need to behave altruistically in a biological sense. What can replace its foundation now that altruism falters? What would compel women—and men, too, if the drugs degenerate their sperm or their sperm count, or in other ways impede their ability to reproduce—to refuse potential addictions? We need to devise a morality that tells us our priorities and provides us with the moral conviction to carry them out. We seek a modern content for altruism to promote biological altruism.

Reflecting on the lessons from sociobiology, Edward Wilson lists the following values as essential for ethics:

• maintaining and preserving the human gene pool;
• maintaining the diversity of this pool; and
• universal human rights.

Let's apply this to the matter of abortion. For instance, does a fetus possess rights, assuming it (she? he?) preserves or diversifies the gene pool as much as any other unborn? Wilson's list hardly helps here. Sociobiolog-ically provided values fail to produce a morality broad enough to apply in most circumstances. Neither do they always choose between competing biological inclinations. We need something more, namely, culture's expan-sion of the values by rational discussion. Wilson provides a faulty list, but at least it starts the conversation.

Most people, Ruse says, participate in each person's social circle because of modern technology. We turn on the TV and watch young women working for $2.28 a day in a sweatshop in Serang, Indonesia, to produce our shoes. The world becomes smaller and smaller. "Here perhaps our technology has out-run our innate moral sentiments. Our animal nature is, in respects, inadequate to deal with today's problems." Programs devised by culture must supplement biological feelings, Ruse adds, because they cover too small a circle. The command to act altruistically issues from biology, but what does "love one another as yourself" mean when dispos-ing of unwanted frozen embryos? Our innate morality hardly guides us in this circumstance. For culture to devise such moral programs starts with "our evolutionary evolved powers of reason and understanding, and . . . applies them to our ultimate biological self-interest."[2] Ruse builds toward the strong love command, despite his rejection of it, and asks about the last value in Wilson's list, "universal human rights": what does our moral oblig-ation to all people entail?

Someone must finally decide the contents of the morality. Spiritual traditions, atheists, ethicists, and others can help society build an adequate ethics by applying the discoveries of such sciences as sociobiology. But a scientific process bows to a political one, commentators contend, "with its reliance on trial and error and on compromise."[3] Wilson wants the West-ern ideal of democratic consensus to determine social controls, and he wants these decisions weighted toward biologically natural behaviors. In 1996, President Clinton signed out of existence an altruistic tradition of unilateral care for poor mothers and their children. Perhaps this welfare program exemplified biological altruism.

This is too simplistic. You or I or a political compromise may ratio-nally construct a morality, but it may never hold. It needs to grow from within the culture. We create a morality out of our genetic and cultural roots to serve certain biological needs. To succeed, it must reach right down into our social and personal and biological roots. Power provides the key. If the morality emerges from public debate, from where does it draw its strength of conviction? What would provide altruism the power of objective truth so people follow it and thus shoot for biological altruism? The laws of the state and its systems of punishment can force morals on us, but only so far. They have to feel right to us. Following them must emerge from internal strength of conviction. Some people still have to cotton onto the laws stipulating racial equality.

Epigenetic rules perform this job. However, the epigenetic rules that lead to biologically altruistic behavior need social support to hoist their strength over that of opposing inclinations. Belief in divinities and their moral imperatives used to do this. Where can the altruistic epigenetic rules now find their additional support?

Following the Moral Wisdom of Spiritual Traditions and Embracing Modernity

Before the rise of modern science, spiritual traditions retained power both over moral behavior and the explanations of how the universe operates. The perceived clout of the traditions' divinities backed the authority. Now science holds the power of explanation, and the spiritual traditions hold the moral wisdom of society. The leverage that spiritual traditions gave to moral will has dissipated, scattered into hundreds of movements and traditions, each of which claims to preach the right rules for life. The dust jacket of Gopi Krishna's *Kundalini: Empowering Human Evolution,* for instance, says the book holds the secrets of life, the answers to the most pressing questions of our tormented age. Each movement finds power in its beliefs about its divinity or about how it says we should behave. With this power, each supports the epigenetic rules. In general, though, these spiritual traditions and movements shun modernity and conflict with or ignore science, neutralizing some of the power from both. Krishna's book assumes human evolution. But he believes that a life force powers and directs evolution rather than, as science holds, that evolution emerges undirected from mutations and natural selection. How can we activate the power of moral will that compels us to follow the moral wisdom of spiritual traditions and yet embrace modernity?

I ask about the spiritual beliefs of couples who ask me to preside at their weddings. Do they believe in a divinity? Yes, most reply. But you'll never see them in church because, many tell me, their spiritual institution has strayed too much from its original vision. If the couples consider themselves Christian, I ask about their attitude toward Jesus. Jesus was a great man whose life and teachings they try to follow, most respond. I call this attitude a "secular morality" and recognize it as undeveloped and usually passive. Traditional spiritual values have devolved into secular morality. Mass communication and Western society transmit this morality, making it nearly universal. Secular morality emerges as the trans-kin glue that the spiritual traditions with their divinities used to provide.

Power for moral will must come from epigenetic rules bolstered by science and the wisdom in the spiritual traditions. Perhaps science and spiritual traditions need to join forces. Wilson writes,

> The best relation between [spiritual traditions] and science toward which we might aim . . . [is] an uneasy but fruitful alliance. The role of [spiritual traditions] is to codify and put into enduring poetic form the highest moral values of a society consistent with empirical knowledge and to lead in moral reasoning. The role of science is to test remorselessly every conclusion about human nature and to search for the bedrock of ethics by which I mean the material basis of natural law.[4]

"The Bible alone, and the Bible in its entirety, is the written Word of God, our absolute authority, inerrant in the autographs." So declares the Doctrinal Statement of the Interdisciplinary Biblical Research Institute, though it could come from hundreds, perhaps millions, of groups. The feelings of security and assuredness such positions arouse win them the allegiance of masses of people the world over. Dogmatic spiritual traditions and spiritual-like political ideologies draw people in, Wilson says, because they feed off the rigid nature and power of epigenetic rules. Wilson thinks liberal spiritual thought can help defuse this situation because it serves as a buffer between science and the dogmatisms. On the one hand, it competes with fundamentalism by serving the spiritual needs of most people. On the other hand, it discovers new truths from science and strengthens its goals. It learns.

The development of a spiritual yet naturalistic worldview that flows from the epic of evolution might also help. Such a vision, Wilson thinks, competes with orthodox spiritual traditions and will replace them. Further, liberal spiritual thought challenges agnostic and atheistic interpretations of evolution, and this is good. In raising questions about the human mind, it forces secularists to learn and grow. Wilson might appreciate Ralph Burhoe's theories that build on the idea that natural selection is the Divine.

We needn't fear this scientific worldview, Wilson assures his readers. Science constructs ideas as well as analyzes them into pieces. The two operations link. Science explains complex phenomena with simpler ones. It explains the existence of mountains, ocean troughs, earthquakes, fault lines, and volcanoes with its model of an object floating on circulating water, the earth's crust on its magma. But science can also acknowledge the worth of such cultural expressions as art and spiritual tradition; sociobiology sees to this. Science does pursue the spiritual to try to explain it. When it approaches the spiritual, however, Wilson writes, "both will evolve into something new, permitting the capture and the resolution."[5] Each will employ the other.

To join science and spiritual traditions at their deepest levels—as this book sets out to do—revises the traditional image of the Divine. Liberal spiritual thought and the naturalistic worldview both benefit if centered on this model. It inclines us toward moral wisdom and science's explanations. It also encourages us to engage, root, and develop secular morality. When we project this morality onto this divinity, we take from the divinity a secular yet spiritual and moral way of life. We find this projection natural and real. It provides both a guide to altruism and the imperative to follow it, employing the explanatory strength of science to ordain its way of life. It thereby helps us follow altruistic epigenetic rules and, hence, further biological altruism and the survival of humanity.

The image of the Divine assembled by this book promotes and strengthens the mutual relevance of science and spiritual thought. Their union fulfills the aspirations of Wilson: it is naturalistic, drawing from the evolutionary story; it is liberal as opposed to fundamentalist; it is open to science; it develops from the wisdom of spiritual traditions. I hope it also expresses the intuitions of those who comfortably live an integration of the secular-scientific and the spiritual aspects of our culture.

At the same time, this image also exceeds Wilson's desires. Maybe it moves toward the "something new" he envisions by developing an image that unifies the knowledge of science and the wisdom of Western spiritual traditions.

The human species may soon become extinct, anthropologist Sol Katz worries, if it fails to act urgently on its problems. Countries and groups vie with one another. Seldom do they agree over what is right and what is wrong, unless they bow to a military or police force more powerful than either of them. How, then, can the peoples of the world be expected to agree on a global morality? Even if they did, what would attract people to freely follow that morality? Global warming remains on the periphery as an interesting idea; fear of extinction amounts to little. Neither does fear of military might. A global morality backed by both science and the world's spiritual and cultural traditions may suffice, especially with the nearly universal acceptance of secular morality. In our system of spiritual ideas may reside the answer to Katz's plea. Time remains the crucial factor.

Part Four
The Spiritual Quest

CHAPTER EIGHTEEN

SUFFERING AND
EVIL AND HUMANIZATION

Rejecting Suffering and Evil

THE PUZZLE OF EVIL AND SUFFERING has long presented an enigma in Western thought. Throughout history, people have found themselves confronted by two types of evil. When humans act wrongly, they commit moral evil. Terrorists bomb the innocent; the young indulge in drugs; employers fire employees and do so without feeling; children lie to their parents. At times, we all feel the pain of injustice, suffer, and feel a sense of helplessness, and each of us intentionally and unintentionally does wrong. The other sort of evil is natural, as when suffering and destruction happen from natural agencies without human responsibility, whether through earthquakes, hurricanes, genetic diseases, and so on. We all experience both types of evil.

Humans hate suffering. We despise the suffering that evil brings about. In our world, suffering and evil shout at us, and we want an end to them. The Divine also hates evil and suffering, we tend to believe, because the Divine feels our sense of right and wrong, the same morality and sense of justice to which we aspire. The Divine and the good life on the other side of the grave must reject them.

Sorting Out the Divine Involvement

But then we pause to think that the Divine owns the power to do anything, and if the Divine really loves people, the Divine would stop all this misery before it happens. Does the fact that it continues unabated mean the Divine really doesn't love us, or have I said something else awry? If a father forced his ill son to mow the lawn, for instance, you would question the father's love for his boy. John Hick captures this puzzle for traditional Christian thought, and I paraphrase it: If the Divine is perfectly loving, the Divine must wish to abolish all evil; and if the Divine is all-powerful, the Divine must be able to abolish all evil. But evil exists; therefore, the Divine can't be both all-powerful and perfectly loving.[1]

Must the Divine be perfectly loving? Scriptures, the Hebrew scriptures, for example, find the Divine portrayed as meting out, rather than alleviating, suffering:

> Moses then said, "These are the words of the Lord: 'At midnight I will go out among the Egyptians. Every first-born creature in the land of Egypt shall die: the first-born of Pharaoh who sits on his throne, the first-born of the slave-girl at the handmill, and all the first-born of the cattle. All Egypt will send up a great cry of anguish, a cry the like of which has never been heard before, nor ever will be again.'" (Exod. 11:4-6 NEB)

From the Christian scriptures:

> Then another angel, a third, followed them, crying with a loud voice, "Those who worship the beast and its image, and receive a mark on their foreheads or on their hands, they will also drink the wine of God's wrath, poured unmixed into the cup of God's anger, and they will be tormented with fire and sulfur in the presence of the holy angels and in the presence of the Lamb. And the smoke of their torment goes up forever and ever. There is no rest day or night for those who worship the beast and its image and for anyone who receives the mark of its name." (Rev. 14:9-12 NTP)

To moderns, these texts seem anti-Judaic and anti-Christian. The Divine of these traditions, who stands for high ideals of loving and humanization, can't will that suffering. The Divine acts consistently—consistently loving—we believe.

But suffering and evil persist. So we try to excuse the Divine. Even if the reasons lie within a scheme or plan beyond our understanding, we think the Divine hates evil and suffering and in the long run will stop them. Some blame the evil and suffering on Satan and think the Divine relieved this condition of the fallen universe by atonement through Jesus' sacrifice. To escape this condition, others believe, we need only die: the

Divine will mollify it in the last judgment and in heaven. Perhaps the father who forced his son to mow the lawn thought only of the boy's long-term good; as an adult, sometimes you have to roll out of bed and show up at work even when you feel out of sorts.

Calling upon Free Will

In another tactic, spiritual traditions call upon free will to help. Arthur Peacocke writes of sin as falling short of what the Divine intends for us. Perhaps the boy felt ill because he partied hard the night before. As free and self-conscious beings, we can position ourselves at the center of our individual and social lives, opposing the Divine's purpose. We then fail to reach our full potential, an image of the Divine. Peacocke believes it more important for us to keep this freedom to choose our own destiny than for the Divine to prevent any suffering or evil from happening.

When Peacocke discusses natural evil, he turns to evolution. Pain and suffering and the random elements in the universe must exist for the universe to evolve. The universe owes every moment of its existence to the work of the Divine, yet natural evil must exist for self-conscious and free beings to emerge. John Polkinghorne echoes Peacocke: the Divine allows the universe to carry on under its laws. It starts with the potential to evolve and even produce human beings, and the processes of the universe explore and move toward realizing this potential. Yet, the systems are also free to go wrong, and natural evil arises from this risk. As a result, the freewill defense expands into the free-process defense. Polkinghorne and Peacocke believe it more important for the universe to keep the freedom to work out its own destiny according to its laws than for the Divine to prevent any suffering or evil from happening.

The chief Christian models for solutions to the puzzle of evil and suffering are the Augustinian and the Irenaean. The Augustinian model describes humans as created perfect. Then they destroyed their perfection for a reason that now escapes us and "plunged into sin and misery." In contrast, the Irenaean model says the Divine created humanity imperfect and immature. People will grow and develop morally, finally to reach the perfection the Divine intends for them. The fall (the loss of innocence and divine protection when Adam ate the forbidden fruit in the Garden of Eden) marks "an understandable lapse due to weakness and immaturity." In this view, the universe comprises a mixture of good and evil, which the Divine appointed as the setting for humanity's development toward perfection.[2]

Suffering and evil persist despite all our attempts—even the freewill and free-process defenses, Augustinian and Irenaean—to let the Divine off the hook. The experience of suffering confronts us too powerfully and the emphasis on the goodness of the Divine too persistently. Like trying to comfort a parent who has lost a child, the question eclipses academic curiosity. The child remains dead. Acts may help more than words. Other forces than inquisitiveness cause the cross-examination of the Divine, and these overpower all answers.

Exploring the Passion over the Puzzle of Suffering and Evil

Let's explore these deep passions of pain and suffering. What generates such strong feelings and fervent questions of the Divine? Biology. Biology generates these feelings.

• We suffer because we're biological beings. We give weight to suffering and evil because we evolved to do so. Biology encourages us to want a world free from suffering and evil—we require this for our survival.

• Biology also promotes belief in a power beyond the universe who cares for us, as the last chapters discussed. The all-loving nature of the Divine arises because the Western belief system includes it, and biology had Westerners project this worldview onto the Divine. Through its epigenetic rules, biology inclines us to believe in and feel the truth of an omnipotent, absolutely moral, all-loving, and transcendent deity.

For the sake of our survival—the strongest drive in us—biology cries out through us for our release from suffering and evil. For the sake of our survival, biology also creates through us an all-loving, all-powerful divinity. The puzzle of evil and suffering and the power in its tension, therefore, arise from tactics biological evolution instilled in us. It grasps a dichotomy in human nature.

Sociobiology explains the passion over the puzzle of evil and suffering, but it only says why people feel this way, not how to change it. The crying out continues, and the intellectual puzzle of why the Divine allows evil and suffering to happen still waits for a solution. You might be angry with your spouse for totaling the car, but you still need to work out alternative travel, insurance, legal matters, medical care, and so on, in a calm and reasonable way.

The question over evil and suffering assumes the Divine loves us in the same way parents love their children: if you love someone, you shield them from evil and suffering as much as possible, especially from extreme cases. So the question supposes that the Divine pursues our highest sense of morality. We project this onto the Divine. The question also assumes the Divine carries the power to stop the suffering and evil. We project this as well.

But the Divine doesn't abide by a morality as we understand the term. If the Divine follows something similar, it exceeds ours in ways probably beyond our imagining. The Divine isn't, as we understand the phrase, perfectly good.

This unseats the first part of Hick's statement of the problem, "If the Divine is perfectly loving, the Divine must wish to abolish all evil." It therefore cancels the conclusion, "The Divine can't be both all-powerful and perfectly loving." Out flies the contradiction and puzzle.

To approach the puzzle this way and to emphasize growth and development (read: evolution) leans toward the Irenaean in Hick's analysis of approaches, though it declines divine status for morality. It also sees the universe as a mixture of good and evil.

All humans at times experience the depths and twistedness of evil and suffering. We feel alien, hurt, separated, depressed, angry. Sons imbibe too much and fall to their deaths from top floors of fraternities. It astounds me to realize why the existence of evil and suffering disturbs me and to see the intellectual puzzle melt in the Divine's amorality. They astound me because the belief that the Divine could remove the negatives is so powerful. When I think further, I see the traditional nature of this divinity ingrained into me by my upbringing. This divinity fails us. For certain, no divinity did or does away with evil and suffering.

Developing Solutions

My wife tells this story:

> A man was fishing in his boat when it sprung a leak and sank. His companion threw him a life preserver, but the man said, "I don't need that. God will save me." His companion swam to shore. A boat appeared and

the captain yelled out that he would throw out a lifeline and pull him in. Again the man answered, "I don't need your help. God will save me." The boat left and in a short time a helicopter appeared. The pilot shouted down that they would throw him a rope and pull him in. Again the man answered, "I don't need that. God will save me." The helicopter left, and the man drowned. When he arrived in heaven, he asked God why he hadn't saved him. God replied, "Not save you! I sent you a life preserver, a boat, and a helicopter, and you refused all of them."

We can help overcome many instances of suffering with technology, ancient or modern, Eastern or Western. And we can develop and empower moral systems that discourage evil acts and actions that cause suffering. We frequently wait around or clamber for freebies from the Divine like children in a scramble for candy thrown in the air. Instead, we should solve the problems ourselves. The Divine evolved us. The Divine also unfolds the means we operate by, both physical and moral. The Divine thus helps us to overcome evil and suffering—and to relieve and remove natural evils (disease, earthquakes)—by the evolution in us of the tools to do so. This leaves solving puzzles to us with our abilities. The Divine fixes our problems by having us fix them.

According to *The American Heritage Dictionary, altruism* is "unselfish concern for the welfare of others; selflessness." We could develop moral systems that encourage altruism and hence promote biological altruism, thus employing our abilities to remove some pain and suffering. To behave this way means to cooperate, act out our concern for others, share, and strive for peace. In short, we would work against oppression and for humanization.

This adds to what we arrived at before. The individuals in a society can play diverse and fulfilling roles toward a goal that unifies them. But even individuals in Nazi Germany functioned this way. My emphasis on wholeness-with-diversity in chapter 12 leads us only so far toward humanization. The sociobiological approach provides what chapter 12 missed: an emphasis on altruism as a moral guideline, wholeness-with-diversity through altruism.

Yet we must feel some pain, because it tells us something's awry and requires attention. My leg hurts because I fell down the stairs and broke it. We must also condone actions that appear evil to some so we can oppose other evils. Should we have fought Nazi Germany, an evil power, and killed millions of people, or not? Grays permeate morality. Though altruism leads as the paramount moral drive, some of its opposite must exist, too. Further, not everything natural is right; while the "is's" inform the "oughts," it remains vague under what conditions we can classify an "is" as also an "ought." It's difficult to know in detail what to do about evil and suffering; we need to work out actions appropriate for various types of circumstances. We need a morality that inspires altruism in whatever form becomes appropriate. If merit resides in sociobiology, much interesting spiritual and ethical thought and construction awaits.

Polkinghorne wrote in a letter to me, "I am afraid I can't believe that morality is simply a human construction or that the Divine is beyond morality." To remove the perfectly good from the Divine steps away from orthodox Christian thought. I react, "But it doesn't take away the good as a human ideal." From there we proceed.

CHAPTER NINETEEN

CHRISTIAN BELIEF

Speaking in a Christian Context

Through the last eighteen chapters, I have developed a system of spiritual thought that tradition would call a "natural theology." In so doing, I have constructed it independently of a religious orientation. Now I introduce several specifically Christian matters that build on this natural theology. Tradition would call a theology enriched this way "revealed."

Ideas central to Christian theology include "revelation" (a disclosure of divine will or truth, specifically the revelation who is Jesus Christ) and "resurrection" (the rising of Jesus on the third day after his crucifixion, and the rising of the dead). These create Christian theology because they center on Jesus. Many other ideas exist in this system of theories, but these two lie most centrally. Note the absence of "creation," the main subject of the natural theology developed in this book. Christian theology includes it in its base but not in its focus.

You may notice that once again I resort to the words *theology* and *God.* In this chapter, I break my linguistic restriction and revert to traditional words: *God* instead of *Divine,* and *theology* instead of *a system of spiritual ideas.* I speak in a specifically Christian context here and address the words of that tradition. I want to fill these words with fresh meaning drawn from what I have developed so far in this book. With these, I hope to replace the terms of old devotions.

Asking Questions about Revelation

I can't accept Jesus as the revelation of God, as the incarnation, without question, because I need to understand the context of the belief. Contexts hide meanings. The first Americans I saw, for instance, spoke in a language that stumped me; someone else told me those others spoke "Yank."

Hearing a southern drawl like that now makes me realize that a Yank is a Yankee, someone from New England, or at least from the north. In the United States, the word *Yank* possesses a context of meaning that outsiders probably know nothing about; how could outsiders feel the latent hostility here between many northerners and southerners?

The traditional Christian context for "Jesus is the revelation of God" hides many beliefs: What is God? Why would God want to become a human being? What does it mean for Jesus to be both God and human? Traditional answers to such questions and traditional understandings of the phrase irk me. I want to feel true to the new system of spiritual ideas I generate. I accept *revelation* and *resurrection* as terms to explain, but I refuse to import orthodox interpretations along with them. New wine requires new wineskins.

So I rescind a phrase in the definition of *revelation* ("a disclosure of divine will or truth, specifically the revelation who is Jesus Christ"). Specifically, I reject the existence of divine will or truth, as these project too personal a nature onto God. By *revelation,* rather, I refer to the disclosure of specific truths about human nature. God is the subuniverse, the unfolder of everything, so self-revelation by God to humanity would disclose specific truths about human nature within the context of all that exists, existed, and will exist.

To provide a context that helps us understand the nature of the revelation and thus what God would self-reveal, I ask: Why would God self-reveal?

Religion started as tribal and animistic, where people saw nature as shot through with power that they could influence by ritual. It then expanded its pools of adherents so that a religion applied to larger tribal groups. This stage saw the advent of religions such as those of the Greeks and Vedic Indians, who believed in a high god or supreme being beyond animistic and tribal gods and powers. The personalities of the gods became more pronounced, with speculation about the ultimate nature of reality and personal devotion moving to the fore. Then, starting about the sixth century B.C.E., what we now know as the major religions began to form. The teachings of Confucius and Lao-tzu in China expressed new mystical and ethical insights. The Upanishads and the Buddha gave birth to a new form of the religious quest in India. And in Israel, Moses and other prophets preached a monotheism that stood against the beliefs of surrounding tribes. Jesus and Christianity emerged out of Judaism. The

Judaic and Christian traditions also influenced the prophet of Islam, Mohammed. At this stage, religion became international across cultures, tied to the spread of settled agricultural societies and related to the development of trans-kin altruism beyond the boundaries of related tribes. Religions faced natural selection of a cultural type and few passed the test of time. The Hellenistic religion of the Greeks and Romans failed to survive, for instance. Religion participated in the evolutionary story of humanity and played an essential role in forging cohesion among a variety of cultures.

So to the question "Why would God self-reveal?" I respond: at that juncture of human (cultural) evolution or development, key figures disclosed truths about human nature, and around them movements emerged that grew to change the course of history. Self-revelations of God became necessary about 2,000 years ago (give or take 600 years) for the next natural stage in the evolution of cultures to occur. Christianity, one of the transnational religions, obviously did survive. Jesus founded it, or at least formed the focus of its founders, Paul, Peter, and other of Jesus' followers, and of all its adherents.

What form might a self-revelation of God require? Why a person?

A cat named Fred once lived with me in a house beside a busy road. One evening, Fred failed to return home, an unusual event. I had noticed her on the other side of the road on occasions, so I feared the worst. After an hour searching on both sides of the road and calling her name innumerable times, I finally heard a weak and plaintive meow from under a shrub. A vehicle had hit her. She survived with operations and much care, but I would still see her on the other side when she was better. How could I warn her of the dangers and tell her not to cross? All my talking and affection achieved no ounce of good. I would have to become a cat and convince her in cat talk of the perils of the road. Similarly, so my Christmas sermon would explain, God had to become a human being to tell us about the nature of God. For a message to communicate to us at all human levels, it must come from a person.

A revelation as a person can communicate to us at all personal levels, but does that establish it as a revelation of God? Every person, everything we know of—even inspiration—is a self-revelation of God, because all unfold from God the subuniverse. God self-reveals in Jesus and all the other prophets.

Why should we devote ourselves to Jesus, then, rather than to any of the other great teachers? Because we choose to. Many people raise George Washington high on a pedestal, while no one that I know elevates King George III of Britain, monarch during the revolutionary wars. Many significant individuals have lived and many live now. Some of these we choose as more central. The Jesus movement, along with several others, stands the test of time; it formed and still forms much of the backbone of Western civilization. I choose it because of this and my desire to continue as an active part of this culture.

Jesus Christ is a revelation of God that I select to follow and believe. To choose this way is to leap in faith, to commit myself, to become a Christian. Rather than existing as a theoretical assumption, accepting Jesus as the central revelation of God means to believe in the centrality of what Jesus said and showed and to try to follow its implications for my life. If I became a cat and Fred accepted me as the provider of meaning for her life, she would stop crossing the road.

Why should God want to self-reveal? I wish to communicate with you; the human race wishes to communicate with extraterrestrials; God wishes to communicate with us. Arthur Peacocke follows this route when he tries to understand why God wants to self-reveal. A personal God wants to disclose Godself and would do so in a personal form. But as I have overemphasized already, I shy away from using words like *moral* or *purpose* of God. The properties of God may so far exceed our human ones that they cease to resemble ours, and to claim the personal likeness of God dangerously projects human qualities onto something we can as yet hardly fathom. Thus, unlike Peacocke, I hesitate to personify God and decline to talk of God as "wanting" to do anything, including self-reveal. Rather, God self-communicates as a natural part or progression of the evolution of culture; God unfolds it all and this self-communication follows that pattern. Western culture directed a portion of that evolution when it chose Jesus as its focus.

What does Jesus reveal? Jesus' revelation focuses on the experiences that touch the heart of our lives: love (altruism), inspiration, hope, suffering, and death, for instance. It tells us, first, that we should interpret and approach life in terms of altruistic love for each other, in terms of actions for the disadvantaged. The Christian model provides a general approach for how to live this out in everyday life. As Jesus tells the rich young man, "Sell all that you own and distribute the money to the poor" (Luke 18:22 JB). Or as James writes, "Coming to the help of orphans and widows when they need it" (James 1:27 JB).

Michael Ruse objects to the strong love command from his sociobiological point of view because he says it runs against what we would consider biologically natural. To forgive someone 490 times, you may remember him saying, is irresponsible. Like a biblical fundamentalist, he interprets the commission as something we could literally achieve. The Christian love command, however, instead creates an ideal that no one will fulfill to the letter. To achieve a lofty goal, we must aim for something higher still.

Around us, our goals seem low. We fill our cities with filth. As individuals, corporations, government bodies, we dump garbage beside roads, into waterways—everywhere. Such selfish actions need to balance with our need for survival, which in turn requires the survival of our environment. For the human species to continue and for the repair and maintenance of the social and cosmic environment, we must limit our self-centered qualities (also necessary for survival), because their overuse threatens the survival of our species. We must hold them in balance. Jesus' examples of love pushes altruism and, therefore, biological altruism; it can achieve this balance.

Biological altruism generates altruism, the attitude that God (our biological rootedness) prompts us to hold toward one another. Thus this self-revelation by God—our perception of the life and teachings of the man Jesus Christ—fits with the laws of nature. God works this way. The second aspect of Jesus' revelation aligns altruism with God, the whole that is the subuniverse. Christians like to say, "God is love" (1 John 4:8 JB).

Asking Questions about Resurrection

After "revelation," "resurrection" forms the second plank of Christian belief. Jesus died and Joseph of Arimathea placed his body in a tomb. After a couple of days, according to Luke's Gospel,

- women went into the tomb to prepare his body for full burial, but he had disappeared;
- Peter also found the tomb empty;
- Jesus walked and ate with two of his disciples on the road to a village called Emmaus; and

- Jesus appeared to his eleven apostles, talked with them, showed them his hands and feet, and ate with them.

Was he only a force in his followers' minds, like George Washington in the minds of living American idealists? Did he have a physical body? Was he a ghost like Casper? Something happened to the person of Jesus after his death by which he continued in some form of life and interacted with others. This seems an indisputable base. But what did happen and what it means aroused and continues to arouse controversy. What happened historically, and where does interpretation begin? Like John Polkinghorne, I can't say what form the resurrection of Jesus took, but it's the fact of the resurrection that is more important than what emerged from the tomb between Jesus' temporary burial and the arrival of the women. The resurrection dominates the empty tomb.

What does Jesus' resurrection signify? The full impact of God's self-revelation requires the event that we call the "resurrection." It helps interpret the life of Jesus and hence its meaning for us. Conversely, to understand or recognize his resurrection requires an understanding of who he was and is. The meaning of this resurrection, something that happened to Jesus, also calls upon the revelation. Logically, for Christians, the two are inseparable.

The fall of the Berlin Wall marked the beginning of the decline of Soviet communism in Eastern Europe. It signifies for some people that liberal democracy and its fellow consumerism will succeed in the end. Though the analogy is limited, the resurrection similarly signifies to those of the Christian faith that Jesus' altruistic way will succeed in the end. Altruism will win, biological altruism will win, and humanity will survive. Evil and suffering continue, but as Jesus suffered pain and humiliation and God left them alone, so the best will happen through them and in spite of them. Altruism exists to the end, and the universe becomes more whole: this is the hope in the alternative way of Jesus. Resurrection, therefore, emerges from biology and culture to become significant and important for us in the evolutionary survival of the human species.

For Christians, Jesus indicates the essential properties for being human. His resurrection establishes hope in the success of the way of altruism: that this way will beat rival ways of life. It also grounds a second hope, that there is more to our lives than bodily death. Jesus' resurrection suggests our resurrection, our life after death. If he can do it, so can we.

But what is this "more"? What is eternal life? The natural theology developed in these pages emphasizes a wholeness in which entities retain their individuality but unite within the whole, which in turn affects the behavior of each. Similarly, all time exists in the subuniverse. It contains no divisions: I always existed there and I always will. The subuniverse expressed me in its unfolding that is me now, on earth, my body, mind, spirit, and experiences. I never leave the wholeness, and while on earth I only weakly perceive it. Each of us will experience an individualized afterlife within the whole into which we enfold. Each of us will affect all other events, even more than we recognize now. Each of us will implicitly contain all other persons, everything, and the whole. All of us will, all of the time, constantly connect with all other parts of reality, past, present, and all the possibilities for the future. Life after death reels in wholeness. We will then explore the endless possibilities of the subuniverse. Resurrected Jesus exemplifies this universal wholeness.

Altruistic behavior by humans marks the best way of life within the web of connections that is the universe-as-a-whole. It helps hold the structure of the web together, like an elastic glue, which in turn centers life after death. Thus, the two hopes the resurrection of Jesus imparts—eternal life and the success of altruism—relate closely.

This thumbnail theology centers on the resurrection, on the wholeness that the subuniverse unfolds. I can say

- wholeness comes first (the subuniverse), which leads to
- evolution and the evolution of humans, which lead to
- the revelation of Jesus (altruism), which leads to
- Jesus' resurrection, which signifies
- life after death, which is a form of
- wholeness.

The resurrection speaks of our life's mystery in the greatest of all mysteries: wholeness, God as the subuniverse.

Note the partial nature of resurrection. It neglects our physical bodies, for instance. Knowing little about wholes, I remain mum about the details of my resurrected state.

Developing Starting Points

> "All right," said the Cat; and this time it vanished quite slowly, begin-
> ning with the end of the tail, and ending with the grin, which remained
> some time after the rest of it had gone.
> "Well! I've often seen a cat without a grin," thought Alice; "but a grin
> without a cat! It's the most curious thing I ever saw in all my life!"[1]

Our smiles continue on. I lay aside the question of what of me or what of
the universe resurrects and what fails to, why that occurs, and how it adds
up. I lay aside some matters, but I answer other more important ones. I try
with them to create the rudiments of a Christian theology. I want the
Christian emphasis because I think the religion for Western society needs
to root itself in Judaism and Christianity, given its cultural heritage. To
develop along Christian lines, the theology must center on Jesus the
Christ, and it must relate positively to the Christian tradition. But Western culture grows in an international context, which means it involves
more and more other cultural traditions. The highly effective motivator,
Stephen Covey, author of *The 7 Habits of Highly Effective People*, is a Mor-
mon who quotes liberally from a variety of wisdom traditions: ancient
Egyptian, Native American, Greek, Hebrew, Christian, Chinese, Indian.
This theology should do likewise, adding the wisdom of other traditions to
what it receives from its predecessors.

On the other hand, I developed this theology via traditional categories
and topics because I think that in this way communicates with the Christian tradition. I need to address standard themes. Yet, what I write fails the
truth tests of creationists and some other Christians. Theologies and
philosophies of the past and present provide some insight into models of
the whole that may prove useful, but they provide only suggestions not
definitive truths. Our tradition ideally addresses each new generation in
each of its settings. I waver about what I want to say apropos God and I
hesitate to jump on any boat of tradition and throw on board beliefs that
invoke good feelings. And, of course, what I write here only starts the
development of the ideas along Christian lines. I know I only skim the surface, but at least I have a place to start. I must load something on my ship
to introduce a Christian perspective.

If I sailed with or toward the beliefs of other spiritual traditions, I
would similarly load key ideas of theirs instead of the Christian ones

I employ now. I develop a system of natural spiritual ideas with which people can interpret their traditions to themselves. How adequately it performs this role remains for further work.

The Christian theology this chapter introduces centers on Jesus. It applies the God idea this book develops, and it attempts honesty about my experience. You may find this last criterion intangible; sometimes I find it so, as well. I listen to all my experiences, I feel them, and then I compare them with what I know and sense about my life and about modernity and the way culture moves through the ages. Sometimes this procedure lasts a long time. Spiritual ideas, mine included, can miss their target. I refuse to follow tradition blindly or traipse behind a fundamentalism—biblical or traditional. Rather, I develop starting points, perhaps startling points, that strive for honesty in tradition and experience.

CHAPTER TWENTY

SCIENTIFIC AND SPIRITUAL THOUGHT

OPEN TO CHANGE

Developing an Understanding of the Divine

A RECENT ARTICLE BY PAT KANE in the *New Statesman* ends with a plea for a

> culture of possibility . . . where the gulf between the humanities and the sciences is filled with a crackling, unpredictable interaction. Not the two cultures, but a third culture: not supernatural—but futurenatural. Maybe not a New Age, or even a New Enlightenment.[1]

The sciences study a manifestation of underlying reality, a spiritual reality that I call "the Divine." To study the physical universe is to study the Divine. The physical world depends on the spiritual as the subsistence of waves depends on that of the ocean. Nothing can exist independent of the Divine. Further, waves form only a tiny part of the ocean; the physical likewise forms only a tiny part of the spiritual, the Divine. "God," John Templeton writes, "is all the things seen and also the vastly greater abundance of things unseen." Perhaps, Templeton suggests, we have as yet discovered at most 1 percent of that reality:

> Maybe no one has ever understood over one percent of "the Divine." Maybe methods of science can double or quadruple our spiritual information. The benefits could be both unexpected and enormous.[2]

This book asks about the Divine of our common belief system and how this divinity interacts with the universe. It constructs a coherent system of spiritual ideas to develop that understanding of the Divine and explores that image within our Western system of beliefs. The methods it employs are empirical. Much of the data and theory it builds upon are drawn from science.

This only starts the exploration. Albert Einstein wrote:

> I maintain that the cosmic religious feeling is the strongest and noblest motive for scientific research. Only those who realize the immense efforts and, above all, the devotion without which pioneer work in theoretical science cannot be achieved are able to grasp the strength of the emotion out of which alone such work, remote as it is from the immediate realities of life, can issue. . . . It is cosmic religious feeling that gives a [person] such strength. A contemporary has said, not unjustly, that in this materialistic world of ours the serious scientific workers are the only profoundly religious people.[3]

Scientific spiritual thinkers likewise toil through failures to understand the spiritual-physical universe with the same painstaking tools as the scientist. Einstein's profoundly spiritual people now include some of those who study the spiritual.

Beds of roses may not greet us, however. The ideas of Galileo Galilei about the cause of tides and his discoveries of the phases of Venus convinced him of the truth of the Copernican theory—that the earth revolves around the sun—against the then-reigning belief that the earth stood still at the center of the universe. Philosophers reacted angrily, scorning his discoveries. They knew that only perfectly spherical bodies exist in the heavens, and that nothing new can appear in them. Aristotle had said so. As a result, church authorities tried Galileo on "grave suspicion of heresy." They compelled him to renounce his ideas, and he spent the rest of his life under house arrest.

If you enact the mutual relevance of science and spiritual thought, you deliver spiritual ideas as suggestions to science, scientific ideas as suggestions to spiritual thought, and you steer the two to interact with and add to each other so they form a more complete understanding. Mutual relevance works out how the Divine acts on and in the universe, and it deploys spiritual explanations to address problems in science. All of this counters the usual belief that spiritual thought is independent of science. The spiritual has its own truth, and science has its own truth. Galileo was guilty of pointing out the conflict between science and spiritual belief and, he noted in his more outrageous defense, the irrelevance of biblical passages to science. The tables now have turned. Today, spiritual tradition believes in the separation, and a minority offer the conjunction of the two. It's the mutual relevance of science and spiritual thought that currently faces the "grave suspicion of heresy." How might we respond?

Advocating the Mutual Relevance of Science and the Spiritual

That the universe behaves consistently is an idea that derives from two spiritual assumptions: that the Divine creates the universe and that the Divine acts consistently. A divinity that behaves consistently will produce a universe doing the same. With such notions, science already depends on and upholds spiritual thought. The idea of mutual relevance only moves this into the open.

Spiritual thought also depends on science. What does the Garden of Eden story—with its cast of Eve, Adam, serpent, apple, and the Divine—mean? You probably avoid the literal understanding with a geographical location for paradise somewhere in the Middle East. More than likely, you see it as a story about good and evil in general and how we humans tend to do wrong. This metaphorical understanding of the Genesis story calls on the tools of biblical criticism developed from scientific, literary, and historical studies. As science already depends on spiritual thought, so spiritual thought already depends on science. Their mutual relevance requires spiritual theories to draw deliberately from and depend on those of science.

But such revelations hardly ease suspicion over mutual relevance. Meteorology fails to explain why the hurricane (described in chapter 1) shifted its course and so avoided the town that many believers live in. The God of the television preacher must have done it, according to the claim of some. His Divine-of-the-gaps explanation of the weather that day in that place invokes divine action to explain something that puzzles established science. The mutual relevance of science and spiritual thought involves the Divine in a similar situation, because it suggests science uses spiritual ideas. A system of spiritual thought motivated by the image of mutual relevance will focus its own demands on science. Should you receive them as skeptically as the preacher's?

The preacher's belief brings with it ideas about his god: He looks out for His believers and cares little for those in the neighboring towns who lost their property, houses, and lives. Divine-of-the-gaps proponents usually differ from the approach of this book, because they start with a full-fledged image of the Divine when they describe how their divinity acts and when they demand the superiority of their explanations over science's. The mutual relevance–oriented system of spiritual ideas doesn't push an already defined divinity, a completed idea of the Divine with all its physical effects, into the universe that science describes. Rather, it develops a basic understanding of the Divine in discussion with both science and spiritual thought. It offers hypotheses to science for perusal and evaluation. So its Divine-of-the-gaps position differs substantially from most others.

What happens if science later rejects one of its theories around which we've molded our spiritual thought? This is the main reason given against a close relationship between spiritual thought and science. What happens if Charles Darwin and evolution fall flat on their faces? Do our spiritual ideas follow? Does such a fate visit those who allow science to affect their understanding of the Divine? We subject the spiritual ideas to rejection and development by science if we introduce them into science or if they draw on its knowledge. A Divine-of-the-gaps position arouses this fear. Changes in scientific theories threaten belief in the everlasting and unchanging nature of the Divine. Some contend that if the Divine never varies, neither do ideas about the Divine.

Beliefs around the Divine and ritual and ecclesiastical traditions frequently reinforce society's power holders, as they did for Galileo's accusers. The political and social forces involving believers shape their ideas. This is also a lesson from the spiritual movement that seeks liberation from oppression: those in power fashion popular beliefs to substantiate their position. Further, the history of science and spiritual thought shows that, though usually unnoticed by believers, spiritual ideas work around changes in science. In 1992, the Roman Catholic Church formally acknowledged its error over Galileo, a position it had implied for many years previously. Spiritual ideas are impermanent constructions subject to science and other social forces.

Opening Ourselves to Change: Adapting Constructively

Spiritual thought isn't unchanging truth that accumulates without backtracking. But is science? Galileo's ideas defied Aristotelian and Ptolemaic science. Quantum physics and relativity build on the inadequacies of Newtonian physics. We now hold to the chemical explanation of fire and discard the belief that the imaginary element phlogiston causes combustion.

Only a couple of decades ago, the big bang theory replaced its rival that thought new matter continuously comes into existence to keep the density of the expanding universe constant. Science does change its mind. What I say about spiritual thought I can also say about science.

The truth of all ideas passes, not just for some of them. You could ask where, if anywhere, you cease uncertainty and claim certitude. How far should the openness venture into your core notions of the Divine or the universe? Do you unconsciously stop before you descend too far? Most of us call off our questions at some point and seek permanence and immunity. We want security. Effective security, though, tracks differently than the permanence of ideas.

Scientific theories are impermanent constructions subject to spiritual ideas and other social forces. Both scientific and spiritual thought move and grow with shifts in human experience and information. That's the way knowledge should be. Whatever else it is, security involves openness to change and the ability to adapt constructively.

Chemotherapy works wonderfully for some people, but not for others. Medicine isn't sure why. Neither does it understand why imaging the dissolution of a tumor or growth can sometimes make it disappear. For certain individuals, meditative techniques work where medical procedures fail. This also eludes science. Why are some colors more restful than others? Why do we find plants and windows essential in a room, though it has adequate lighting and air without them? Why do babies fail if not carried around, cuddled, and talked to, though they receive the proper food and liquid intravenously? The sacrament of communion in the Christian tradition offers bread and wine, not only because they stand for the food and drink our bodies require, but because they symbolize spiritual sustenance. We need both. Without a healthy diet, exercise program, social interactions, and ample time for creative pursuits each day, we quickly lag. We need both science and the spiritual, and, because we are whole beings, we need them to work together, drawing on the strengths of each other to produce a whole system of explanation. The spiritual requires the knowledge of science. Science requires the wisdom of spiritual thought as a spring of ideas to explore.

NOTES

1. Scientific and Spiritual Thought: Their Mutual Relevance

1. Bertrand Russell, *The Free Man's Worship: Philosophical Essays* (Portland, Maine: T. B. Mosher, 1923 [1903]).

2. Robert Wright, "Science, God, and Man," *Time* 140 (28 December 1992): 40.

3. Richard N. Ostling, "Galileo and Other Faithful Scientists," *Time* 140 (28 December 1992): 43.

2. Connection

1. Charles Mann, and Robert Crease. "Interview: John Bell." *Omni* 10 (May 1988): 84–92, 121.

2. F. A. M. Frescura, and Basil J. Hiley, "The Implicate Order, Algebras, and the Spinor," *Foundations of Physics* 10 (February 1980): 8.

3. John F. Clauser, and Abner Shimony, "Bell's Theorem: Experimental Tests and Implications," *Reports on Progress in Physics* 41, no. 4 (1978): 1881.

4. "Passion at a Distance Isn't Very Satisfying," *Discover* 7 (April 1986): 12.

5. Ibid.

6. Jean-Marc Levy-Leblond, "Towards a Proper Quantum Theory (Hints for a Recasting)," in *Quantum Mechanics, a Half Century Later: Colloquium on Fifty Years of Quantum Mechanics,* University Louis Pasteur, 1974, ed. Jose Leite Lopes and Michel Paty (Dordrecht, The Netherlands: D. Reidel Publishing Co., 1977), 188.

3. Separation with Connection

1. David Bohm, "Further Remarks on Order," in *Sketches,* vol. 2 of *Towards a Theoretical Biology,* ed. C. H. Waddington (Chicago: Aldine Publishing Co., 1969), 42.

2. Ibid., 43.

3. David Bohm, and Basil J. Hiley, "Some Remarks on Sarfatti's Proposed Connection between Quantum Phenomena and the Volitional Activity of the Observer-Participator," *Psychoenergetic Systems* 1 (1976): 178.

4. David Bohm, *Wholeness and the Implicate Order* (London: Routledge and Kegan Paul, 1980), 31.

5. Ibid.

6. Mann and Crease, Interview, 86.

7. John Taylor, "Latter Day Gurus of Science," *New Scientist* 109 (30 January 1986): 64.

8. Thomas E. Legere, "Contextual Piece for the Project Demonstrating Excellence: '*A Popular Book—Christianity for a New Age*,'" Ph.D. diss., (1993), 4.

4. The Subuniverse Divine

1. Willem B. Drees, *Beyond the Big Bang: Quantum Cosmologies and God* (La Salle, Ill.: Open Court, 1990), 72.

2. Laurence A. Marschall, "Private Universes," *The Sciences* 31 (March/April 1991): 48.

3. Drees, *Beyond the Big Bang,* 192.

4. Stephen W. Hawking, *A Brief History of Time: From the Big Bang to Black Holes* (New York: Bantam Books, 1988), 174.

5. C. M. Patton and John A. Wheeler, "Is Physics Legislated by Cosmogony?" in *Quantum Gravity: An Oxford Symposium*, ed. C. J. Isham, R. Penrose, and D. W. Sciama (Oxford: Clarendon Press, 1975), 575.

6. Viggo Mortensen, "The Status of the Science-Religion Dialogue," in *Evolution and Creation: A European Perspective*, ed. Svend Andersen and Arthur Peacocke (Aarhus, Denmark: Aarhus University Press, 1987), 197.

6. The Divine Acts?

1. C. W. Rietdijk, *On Waves, Particles and Hidden Variables: A New Approach* (Assen, The Netherlands: Van Gorcum and Co., 1971), 130.

2. Arthur Robert Peacocke, *Theology for a Scientific Age: Being and Becoming—Natural, Divine, and Human,* 2nd ed., (Minneapolis: Fortress Press, 1993), 60–61.

3. Virginia Stem Owens, *And the Trees Clap Their Hands: Faith, Perception, and the New Physics* (Grand Rapids, Mich.: William B. Eerdmans Publishing Co., 1983), 59, 130–31.

7. What Does the Divine Do?

1. Joshua 10:12-13, The Jerusalem Bible.

2. A. P. Elkin, review of *Belief, Language, and Experience,* by R. Needham, *Oceania* 45, no. 1 (1974): 82.

3. James T. Borhek, and Richard F. Curtis, *A Sociology of Belief* (New York: John Wiley and Sons, 1975), 5.

8. The Real Divine

1. John C. Polkinghorne, "The Nature of Physical Reality," *Zygon: Journal of Religion and Science* 26 (June 1991): 222.

2. Richard Moran, "Try Trickle-Up: It May Just Work," *Concord Monitor,* 1 April 1995, B4.

9. Mystery

1. Lewis Carroll, *Alice's Adventures in Wonderland* (London: William Heinemann, 1907), 2–6.

10. The Character of the Divine

1. Plato, *The Republic*, trans. and introd. Desmond Lee, 2nd ed. (London: Penguin Books, 1974), 256–57.

11. Freedom

1. Alfred North Whitehead's term, quoted in John C. Polkinghorne, "A Note on Chaotic Dynamics," *Science and Christian Belief* 1 (October 1989): 126.

12. Liberation and Values

1. John Donne, "Mediation," *Devotions upon Emergent Occasions,* ed. Anthony Raspa (Montreal: McGill/Queen's University Press, 1975), 17

13. Histories of the Universe

1. Thomas Wolfe, *You Can't Go Home Again* (New York: Harper and Brothers, 1940).

2. Richard Dawkins, *The Selfish Gene*, 2nd ed. (Oxford: Oxford University Press, 1989), 192.

14. Sociobiology

1. Eric Bazilian, "One of Us," on compact disc, Joan Osborne, *Relish* (New York: Polygram, 1995).

2. Edward O. Wilson, *On Human Nature* (Cambridge, Mass.: Harvard University Press, 1978), 167.

3. Arthur Robert Peacocke, *God and the New Biology* (San Francisco: Harper and Row, 1986), 111.

15. Does Morality Come from the Divine?

1. Michael Ruse, *The Darwinian Paradigm: Essays on Its History, Philosophy, and Religious Implications* (London: Routledge, 1989), 258, 265.

2. Ibid., 268.

16. Does Morality Come from Biology?

1. Edward O. Wilson, and Thomas M. King, "Religion and Evolutionary Theory," in *Religion, Science, and the Search for Wisdom: Proceedings of a Conference on*

Religion and Science, September 1986, ed. David M. Byers (Washington, D.C.: Bishops' Committee on Human Values, National Conference of Catholic Bishops, 1987), 100, 102.

2. Arthur Robert Peacocke, "Sociobiology and Its Theological Implications," *Zygon: Journal of Religion and Science* 19 (June 1984): 181.

3. Wilson and King, "Religion and Evolutionary Theory," 93.

4. Ibid., 92.

5. John W. Bowker, "The Aeolian Harp: Sociobiology and Human Judgment," *Zygon: Journal of Religion and Science* 15 (September 1980): 329.

6. Michael Ruse, and Edward O. Wilson, "The Evolution of Ethics," *New Scientist* 108 (17 October 1985): 52.

17. Creating a Morality

1. *USA Today,* 8 July 1996.

2. Michael Ruse, "What Can Evolution Tell Us about Ethics?" in *Kooperation und Wettbewerb: Zu Ethik und Biologie Menschlichen Sozialverhaltens,* ed. Hans May, Meinfried Striegnitz, and Philip Hefner, Loccumer Protokolle, vol. 75, 1988 (Rehburg-Loccum, Germany: Evangelische Akademie Loccum, 1989), 219.

3. Bernard D. Davis, "The Importance of Human Individuality for Sociobiology," *Zygon: Journal of Religion and Science* 15 (September 1980): 279.

4. Wilson and King, "Religion and Evolutionary Theory," 89.

5. Edward O. Wilson, "The Relation of Science to Theology," *Zygon: Journal of Religion and Science* 15 (December 1980): 429.

18. Suffering and Evil and Humanization

1. John Hick, *Philosophy of Religion,* 3rd ed. (Englewood Cliffs, N.J.: Prentice Hall, 1983), 40–41.

2. John Hick, *Evil and the God of Love,* 2nd ed. (San Francisco: Harper and Row, 1978), 214–15.

19. Christian Belief

1. Carroll, *Alice's Adventures in Wonderland,* 78.

20. Scientific and Spiritual Thought: Open to Change

1. Pat Kane, "There's Method in the Magic," *New Statesman* (23 August 1996): 27.

2. John Marks Templeton, personal letter, Nassau, The Bahamas, 1995.

3. Albert Einstein, "Religion and Science," *New York Times Magazine* (1930): 1.

BIBLIOGRAPHY

Alexander, Richard D. *The Biology of Moral Systems.* New York: Aldine de Gruyter, 1987.

Alves, Rubem. "On the Eating Habits of Science." In *Faith and Science in an Unjust World.* Vol. 1 of *Plenary Presentations.* Edited by Roger L. Shinn, 41–43. Geneva: World Council of Churches, 1980.

Austin, William H. *The Relevance of Natural Science to Theology.* New York: Barnes and Noble Books, 1976.

Baelz, Peter. "A Christian Perspective on the Biological Scene." *Zygon: Journal of Religion and Science* 19 (June 1984): 209–12.

Barbour, Ian G. *Issues in Science and Religion.* London: SCM Press, 1966.

———. *Religion and Science: Historical and Contemporary Issues.* New York: HarperCollins, 1997.

Barrow, John D., and Frank J. Tipler. *The Anthropic Cosmological Principle.* Oxford: Clarendon Press, 1986.

Belinfante, Frederik Joseph. *A Survey of Hidden Variables Theories.* Oxford: Pergamon Press, 1973.

Bohm, David. *Wholeness and the Implicate Order.* London: Routledge and Kegan Paul, 1980.

———. "Further Remarks on Order." In *Sketches.* Vol. 2 of *Towards a Theoretical Biology.* Edited by C. H. Waddington, 41–60. Chicago: Aldine Publishing Co., 1969.

———. *On Dialogue.* Edited by Lee Nichol. New York: Routledge, 1996.

———. "The Implicate or Enfolded Order: A New Order for Physics." In *Mind in Nature: Essays on the Interface of Science and Philosophy.* Edited by John B. Cobb Jr. and David Ray Griffin, 37–42. Washington, D.C.: University Press of America, 1978.

———. "The Implicate Order: A New Order for Physics." *Process Studies* 8 (Summer 1978): 73–102.

Bohm, David, and Basil Hiley. "Some Remarks on Sarfatti's Proposed Connection between Quantum Phenomena and the Volitional Activity of the Observer-Participator." *Psychoenergetic Systems* 1 (1976): 173–79.

Bohm, David, and Renee Weber. "The Enfolding-Unfolding Universe: A Conversation with David Bohm." *Re-Vision* 1 (Summer/Fall 1978): 24–51.

Bohm, David, and John Welwood. "Issues in Physics, Psychology and Metaphysics: A Conversation." *Journal of Transpersonal Psychology* 12, no. 1 (1980): 25–36.

Bohr, N. "Can a Quantum-Mechanical Description of Physical Reality Be Considered Complete?" *Physical Review* 48 (15 October 1935): 696–702.

Borhek, James T., and Richard F. Curtis. *A Sociology of Belief.* New York: John Wiley and Sons, 1975.

Bower, B. "Gene Tied to Excitable Personality." *Science News* 149 (6 January 1996): 4.

———. "Whole-Brain Interpreter: A Cognitive Neuroscientist Seeks to Make Theoretical Headway among Split Brains." *Science News* 149 (24 February 1996): 124–25.

Bowker, John W. "The Aeolian Harp: Sociobiology and Human Judgment." *Zygon: Journal of Religion and Science* 15 (September 1980): 307–33.

Brown, Warren S., Nancey Murphy, and H. Newton Malony, eds. *Whatever Happened to the Soul? Scientific and Theological Portraits of Human Nature.* Minneapolis: Fortress Press, 1998.

Burhoe, Ralph Wendell. *Toward a Scientific Theology.* Belfast: Christian Journals, 1981.

Campbell, Donald T. "On the Conflicts between Biological and Social Evolution and between Psychology and Moral Tradition." *Zygon: Journal of Religion and Science* 11 (September 1976): 167–208.

Carroll, Lewis. *Alice's Adventures in Wonderland.* London: William Heinemann, 1907.

Clauser, John F., and Abner Shimony. "Bell's Theorem: Experimental Tests and Implications." *Reports on Progress in Physics* 41, no. 4 (1978): 1881–927.

Clayton, Philip. *God and Contemporary Science.* Grand Rapids, Mich.: William B. Eerdmans, 1997.

Covey, Stephen. *The 7 Habits of Highly Effective People: Restoring the Character Ethic.* New York: Simon and Schuster, 1989.

Crease, Robert P., and Charles C. Mann. "Physics for Mystics [review of *Beyond the Quantum,* by Michael Talbot; *Quantum Physics: Illusion or Reality?* by Alastair I. M. Rae; *The Shaky Game: Einstein, Realism, and the Quantum Theory,* by Arthur Fine; and *The Social Relations of Physics, Mysticism, and Mathematics: Studies in Social Structure, Interests, and Ideas,* by Sal Restivo]." *The Sciences* 27 (July/August 1987): 50–57.

Davis, Bernard D. "The Importance of Human Individuality for Sociobiology." *Zygon: Journal of Religion and Science* 15 (September 1980): 275–93.

Dawkins, Richard. *The Selfish Gene.* 2nd ed. Oxford: Oxford University Press, 1989.

"Deadly Air Bags. Key Finding: Two Children Could Be Killed for Every One Saved"; "Delta Engine Hub Was Split in Two Places"; "8,500 New HIV Cases Occur Daily"; "'Epidemic of Violence on the Job' at All-Time High," *USA Today,* 8 July 1996.

Drees, Willem B. *Beyond the Big Bang: Quantum Cosmologies and God.* La Salle, Ill.: Open Court, 1990.

———. *Religion, Science, and Naturalism.* Cambridge: Cambridge University Press, 1996.

Eaves, L. J., H. J. Eysenck, and N. G. Martin. *Genes, Culture, and Personality: An Empirical Approach.* London: Academic Press, 1989.

Einstein, Albert. "Religion and Science." *New York Times Magazine* (1930): 1.

Einstein, Albert, B. Podolsky, and N. Rosen. "Can Quantum-Mechanical Description of Physical Reality be Considered Complete?" *Physical Review* 47, no. 10 (1935): 777–80.

Elkin, A. P. [Review of *Belief, Language, and Experience,* by R. Needham.] *Oceania* 45, no. 1 (1974): 79–83.

Foster, Patricia L., and Jeffrey M. Trimarchi. "Adaptive Reversion of a Frameshift Mutation in Escherichia Coli by Simple Base Deletions in Homopolymeric Runs." *Science* 265 (15 July 1994): 407–409.

Frescura, F. A. M., and Basil J. Hiley. "The Implicate Order, Algebras, and the Spinor." *Foundations of Physics* 10 (February 1980): 7–31.

Friend, Tim. "Violence Linked to Gene Defect: Pleasure Deficit May Be the Spark." *USA Today,* 9 May 1996, 1D.

Gazzinga, Michael S. *Nature's Mind: The Biological Roots of Thinking, Emotions, Sexuality, Language, and Intelligence.* New York: BasicBooks, 1992.

Goodenough, Ursula. *The Sacred Depths of Nature.* New York: Oxford University Press, 1998.

Hawking, Stephen W. *A Brief History of Time: From the Big Bang to Black Holes.* New York: Bantam Books, 1988.

Hefner, Philip. *The Human Factor: Evolution, Culture, and Religion.* Theology and the Sciences. Minneapolis: Fortress Press, 1993.

———. "Is/Ought: A Risky Relationship between Theology and Science." *Zygon: Journal of Religion and Science* 15 (December 1980): 377–95.

———. "Sociobiology, Ethics, and Theology." *Zygon: Journal of Religion and Science* 19 (June 1984): 185–207.

———. "Survival as a Human Value." *Zygon: Journal of Religion and Science* 15 (June 1980): 203–12.

Helliwell, T. M., and D. A. Konkowski. "Causality Paradoxes and Nonparadoxes: Classical Superluminal Signals and Quantum Measurements." *American Journal of Physics* 51, no. 11 (1983): 996–1003.

Hick, John. *Evil and the God of Love.* 2nd ed. San Francisco: Harper and Row, 1978.

———. *Philosophy of Religion.* 3rd ed. Englewood Cliffs, N.J.: Prentice Hall, 1983.

Hiley, Basil J. "Cosmology, EPR Correlations, and Separability." In *Bell's Theorem, Quantum Theory, and Conceptions of the Universe.* Edited by Menas Kafatos, 181–90. Dordrecht, The Netherlands: Kluwer Academic Publishers, 1989.

Hinde, Robert A. *Why Gods Persist: A Scientific Approach to Religion*. New York: Routledge, 1999.

Holmes, Bob. "Chimps Rise above Law of the Jungle." *New Scientist* 149 (17 February 1996): 10.

Irons, William. "How Did Morality Evolve?" *Zygon: Journal of Religion and Science* 26 (March 1991): 49–89.

Kafatos, Menas, and Robert Nadeau. *The Conscious Universe: Part and Whole in Modern Physical Theory*. New York: Springer-Verlag, 1990.

Kane, Pat. "There's Method in the Magic." *New Statesman* (23 August 1996): 24–27.

Katz, Sol. "Toward a New Concept of Global Morality." Paper presented at the conference "Evolution and Moral Norms: Interdisciplinary Perspectives on the Possibility of Ethics." Loccum, Germany: Evangelical Academy Loccum, 1989.

Kauffman, Stuart. *At Home in the Universe: The Search for the Laws of Self-Organization and Complexity*. New York: Oxford University Press, 1995.

Krishna, Gopi. *Kundalini: Empowering Human Evolution*. St. Paul: Paragon House, 1996.

Larson, David B., and Susan S. Larson. The Forgotten Factor in Physical and Mental Health: What Does the Research Show? An Independent Study Seminar. Unpublished manuscript, 1994.

Legere, Thomas E. "Contextual Piece for the Project Demonstrating Excellence: 'A Popular Book—Christianity for a New Age.'" Ph.D. diss., 1993. Photocopy.

Levy-Leblond, Jean-Marc. "Towards a Proper Quantum Theory (Hints for a Recasting)." In *Quantum Mechanics, a Half Century Later: Colloquium on Fifty Years of Quantum Mechanics,* University Louis Pasteur, 1974. Edited by Jose Leite Lopes and Michel Paty, 171–206. Dordrecht, The Netherlands: D. Reidel Publishing Co., 1977.

Lumsden, Charles J. *Promethean Fire: Reflections on the Origin of Mind*. Cambridge, Mass.: Harvard University Press, 1983.

Lumsden, Charles J., and Edward O. Wilson. *Genes, Mind, and Culture: The Coevolutionary Process*. Cambridge, Mass.: Harvard University Press, 1981.

Mann, Charles, and Robert Crease. "Interview: John Bell." *Omni* 10 (May 1988): 84–92, 121.

Marschall, Laurence A. "Private Universes [review of *Origins: The Lives and Worlds of Modern Cosmologists,* by Alan Lightman and Roberta Brawer, and of *Lonely Hearts of the Cosmos: The Scientific Quest for the Secret of the Universe,* by Dennis Overbye]." *The Sciences* 31 (March/April 1991): 46–51.

Misner, Charles W., Kip S. Thorne, and John Archibald Wheeler. *Gravitation*. San Francisco: W. H. Freeman and Co., 1973.

Moran, Richard. "Try Trickle-Up: It May Just Work." *Concord Monitor,* 1 April
    1995, B4.

Mortensen, Viggo. "The Status of the Science-Religion Dialogue." In *Evolution
    and Creation: A European Perspective.* Edited by Svend Andersen and Arthur
    Peacocke, 192–203. Aarhus, Denmark: Aarhus University Press, 1987.

Newberg, Andrew B., and Eugene d'Aquili. *The Mystical Mind: Probing the Biol-
    ogy of Religious Experience.* Minneapolis: Fortress Press, 1999.

The New Testament and Psalms: An Inclusive Version. New York: Oxford University
    Press, 1995.

Ostling, Richard N. "Galileo and Other Faithful Scientists." *Time* 140 (28 Decem-
    ber 1992): 42–43.

Owens, Virginia Stem. *And the Trees Clap Their Hands: Faith, Perception, and the New
    Physics.* Grand Rapids, Mich.: William B. Eerdmans Publishing Co., 1983.

"Passion at a Distance Isn't Very Satisfying." *Discover* 7 (April 1986): 10–12.

Patton, C. M., and John A. Wheeler. "Is Physics Legislated by Cosmogony?" In
    Quantum Gravity: An Oxford Symposium. Edited by C. J. Isham, R. Penrose,
    and D. W. Sciama, 538–605. Oxford: Clarendon Press, 1975.

Peacocke, Arthur Robert. *Creation and the World of Science: The Bampton Lectures,
    1978.* Bampton Lecture Series. Oxford: Clarendon Press, 1979.

———. *God and the New Biology.* San Francisco: Harper and Row, 1986.

———. *Intimations of Reality: Critical Realism in Science and Religion.* Notre
    Dame, Ind.: University of Notre Dame Press, 1984.

———. *Theology for a Scientific Age: Being and Becoming—Natural, Divine, and
    Human.* Minneapolis: Fortress Press, 1993.

———. "Sociobiology and Its Theological Implications." *Zygon: Journal of Reli-
    gion and Science* 19 (June 1984): 171–84.

Peat, F. David. *Synchronicity: The Bridge between Matter and Mind.* New York: Ban-
    tam Books, 1987.

Penrose, Roger. *Shadows of the Mind: A Search for the Missing Science of Conscious-
    ness.* Oxford: Oxford University Press, 1994.

Peters, Karl E. "Scientific Theology and Spirituality." Paper presented at the *Zygon:
    Journal of Religion and Science* and the Chicago Center for Religion and Sci-
    ence 1993 Templeton Symposium, "Science and Religion: Two Ways of Expe-
    riencing and Interpreting the World," Chicago, 1993.

———. "Evolutionary Naturalism: Survival as a Value." *Zygon: Journal of Religion
    and Science* 15 (June 1980): 213–222.

Peters, Ted. "David Bohm, Postmodernism, and the Divine." *Zygon: Journal of
    Religion and Science* 20 (June 1985): 193–217.

———. Review of *A Brief History of Time: From the Big Bang to Black Holes,* by
    Stephen W. Hawking. *Christian Century* 105 (18–25 May 1988): 513–14.

Plato. *The Republic.* Translated and with an introduction by Desmond Lee. 2nd ed. London: Penguin Books, 1974.

Plomin, Robert. "The Role of Inheritance in Behavior." *Science* 248 (13 April 1990): 183–88.

Polkinghorne, John C. *Science and Providence: God's Interaction with the World.* New Science Library. Boston: Shambhala, 1989.

———. *Serious Talk: Science and Religion in Dialogue.* Valley Forge, Penn.: Trinity Press International, 1995.

———. "The Nature of Physical Reality." *Zygon: Journal of Religion and Science* 26 (June 1991): 221–36.

———. "A Note on Chaotic Dynamics." *Science and Christian Belief* 1 (October 1989): 123–27.

Pribram, K. H. "The Implicate Brain." In *Quantum Implications: Essays in Honour of David Bohm,* edited by Basil J. Hiley and F. David Peat, 365–71. London: Routledge and Kegan Paul, 1987.

Prigogine, Ilya. *From Being to Becoming: Time and Complexity in the Physical Sciences.* San Francisco: W. H. Freeman and Co., 1980.

Ridley, Matthew. *The Origins of Virtue: Human Instincts and the Evolution of Cooperation.* New York: Viking Press, 1997.

Rietdijk, C. W. *On Waves, Particles, and Hidden Variables: A New Approach.* Assen, The Netherlands: Van Gorcum and Co., 1971.

Rottschaefer, William A. *The Biology and Psychology of Moral Agency.* New York: Cambridge University Press, 1998.

Ruse, Michael. *The Darwinian Paradigm: Essays on Its History, Philosophy, and Religious Implications.* London: Routledge, 1989.

———. "What Can Evolution Tell Us about Ethics?" In *Kooperation und Wettbewerb: Zu Ethik und Biologie Menschlichen Sozialverhaltens.* Edited by Hans May, Meinfried Striegnitz, and Philip Hefner, 203–25. Loccumer Protokolle, vol. 75, 1988. Rehburg-Loccum, Germany: Evangelische Akademie Loccum, 1989.

Ruse, Michael, and Edward O. Wilson. "The Evolution of Ethics." *New Scientist* 108 (17 October 1985): 50–52.

Russell, Bertrand. *The Free Man's Worship: Philosophical Essays.* Portland, Maine: T. B. Mosher, 1923 [1903].

Russell, Robert John. "The Physics of David Bohm and Its Relevance to Philosophy and Theology." *Zygon: Journal of Religion and Science* 20 (June 1985): 135–58.

Sarfatti, Jack. "The Physical Roots of Consciousness." In *The Roots of Consciousness: Psychic Liberation through History, Science, and Experience.* Edited by Jeffrey Mishlove, 279–93. New York: Random House, 1975.

Sharpe, Kevin J. *David Bohm's World: New Physics and New Religion.* Lewisburg, Penn.: Bucknell University Press, 1993.

———. *From Science to an Adequate Mythology.* Auckland: Interface Press, 1984.

Sober, Elliot, and David Sloan Wilson. *Unto Others: The Evolution and Psychology of Unselfish Behavior.* Cambridge, Mass.: Harvard University Press, 1998.

Smolin, Lee. "Did the Universe Evolve?" *Classical and Quantum Gravity* 9 (January 1992): 173–91.

———. *The Life of the Cosmos.* New York: Oxford University Press, 1997.

Sugg, Diana K. "Brain-Injured Patients Sue for Place to Go Home To." *Sunday Monitor,* 28 April 1996, E7.

Taylor, John. "Latter Day Gurus of Science [review of *Unfolding Meaning,* by David Bohm]." *New Scientist* 109 (30 January 1986): 64–65.

Templeton, John Marks. Personal letter. Nassau, The Bahamas, 1995.

Templeton, John Marks, and Robert Herrmann. *Is God the Only Reality? Science Points to a Deeper Meaning of the Universe.* New York: Continuum Publishing Co., 1994.

Wilson, Edward O. *On Human Nature.* Cambridge, Mass.: Harvard University Press, 1978.

———. "The Relation of Science to Theology." *Zygon: Journal of Religion and Science* 15 (December 1980): 425–34.

Wilson, Edward O., and Thomas M. King. "Religion and Evolutionary Theory." In *Religion, Science, and the Search for Wisdom: Proceedings of a Conference on Religion and Science,* September 1986. Edited by David M. Byers, 81–102. Washington, D.C.: Bishops' Committee on Human Values, National Conference of Catholic Bishops, 1987.

Wolfe, Thomas. *You Can't Go Home Again.* New York: Harper and Brothers, 1940.

Wright, Robert. "Science, God, and Man." *Time* 140 (28 December 1992): 38–44.

INDEX

afterlife, 37, 130, 147–49, 154–61

altruism, 10, 116–28, 137–42, 152–53, 157–60; biological, 116–20, 125–27, 130, 133–34, 138–42, 152, 158–59; reciprocal, 116–17, 126; trans-kin, 117, 120, 126, 141, 156

anthropic principle, 8, 34–35, 42–46

atheism, 42–43, 55, 59, 128, 139, 142

belief system, 45, 56–70, 163–64

big bang, 6–12, 33–43, 57, 66, 69, 106, 110, 113, 154, 167

biochemicals, 8–13, 87, 118, 123

biology, 87, 96, 107–12, 115, 118–19, 121–39, 142, 150, 158–59

Bohm, David, 15–25, 28–30, 34, 37, 39, 53, 72, 76–78, 83, 105–108, 113, 168, 172–74, 176

brain, 7, 9, 14, 51–53, 61–64, 78, 85, 90, 97–98, 119, 130

chance, 9–11, 48–50, 128

chaos, 60, 102, 107

Christianity, xi, 67, 69, 73–74, 113, 125–27, 131, 137–38, 141, 148–49, 153–59, 161–62, 167

complexity, 8–9, 26, 106–21, 142

consciousness, 29, 61, 67, 82–87, 90–91, 109, 119, 123, 149, 167

consistency, 96, 99, 131, 148, 163–65

creativity, 63–65, 73, 97–98

culture, 116, 119–123, 127–35, 139, 142–143, 157–63

determinism, 48–50, 56, 71, 96, 99–100, 123, 135

Divine, the, projections onto, 9–13, 38, 70–71, 81–90, 99, 114–15, 127–29, 138, 150–57; reality of, x–xi, 3–4, 7, 14, 27, 42–43, 56, 66–72, 75; relationship with universe, 7–13, 16, 23, 28, 38–43, 47–50, 54–59, 63–75, 79–80, 89–91, 95–100, 113, 121, 132, 141–42, 148, 150, 163–66

downward action, 51–54, 65–68

emotions, 9–12, 51, 67, 82, 86, 117–25, 134, 138–41, 147, 150, 161–64

energy, 19, 33, 47–48, 53–56, 68–69, 97, 106–112

environment, 101–103, 135, 158

epigenetic rules, 117–18, 121–22, 125–27, 133–34, 140–42, 150

evil, 71, 113, 123, 147–53, 159, 165

evolution, 3–12, 42–45, 57, 69, 85–88, 96, 102, 109–23, 126–36, 139–42, 149–52, 156–60; socio-cultural, 111, 120, 127–28, 142, 156–57, 166

existence, provision of, 24, 33–34, 36–46, 57, 63, 66, 69–75, 79, 95, 98–101, 106

fundamentalism, 3, 38, 68–69, 141–42, 158, 161–62

genes, 8–11, 48, 87, 96, 99, 109–11, 115–24, 128, 131–35, 138–39, 147

Hawking, Stephen, 8, 34–37, 41–46, 169, 174, 176

hope, 90, 157, 159

information, 18, 21, 23, 67–68, 97, 117, 167

insight, 64, 75–78, 156–57

Jesus, 39, 141, 148, 154–62

Judaism, 74, 137, 148, 155–56, 161

justice, 3, 10, 12, 45, 101–103, 119,
  123–26, 138–39, 147–48, 152,
  157–58, 166

language, 24–25, 40, 59, 76, 86–87;
  religious, ix–x

laws, 140; natural, 9–12, 33–48, 52,
  54, 57–58, 66, 70, 74, 95–99, 106,
  108, 112, 141, 149, 158

levels, 28–29, 39, 49–54, 67–68, 74,
  97, 107, 112–13, 116, 121, 130,
  156

locality, 16–20, 106–109, 112–14

logic, 35–46, 59, 70–71

love, 8–11, 81, 88, 103, 125–26, 137,
  139, 147–51, 157–58

meaning, 1, 8–11, 30, 34, 45, 59,
  86–88, 105, 112–15, 122–25,
  130–31, 134–35, 159

mind, 7–11, 24, 26, 29–30, 39,
  44–45, 51, 53, 61–64, 67, 72, 76,
  78, 85, 96–102, 114, 117–19, 122,
  130, 134, 138–39, 142, 159–60

morality, 3, 6, 9–10, 13, 30, 73, 80,
  88, 104–105, 113–43, 147–53, 157

mystery, 12, 19, 36–37, 72–80, 160

nonlocality, 16–22, 26, 28–30, 53–65,
  86, 90, 97, 106–109, 112–14

order, of our experience (explicate),
  24–27, 30, 54, 76, 113; underlying
  (implicate), 20–30, 34, 37, 39, 54,
  113, 163

Peacocke, Arthur, 67, 73, 86, 103,
  121–22, 131, 149, 157, 169–71,
  176

physics, classical (separateness), 14–29,
  35–36, 70, 75–80, 95, 100,
  104–109, 112–14, 123–24, 132,
  166; quantum, 11, 15–23, 26–29,
  33–38, 42, 48–50, 53, 61–62,
  74–77, 97, 106–109, 112, 166–67

Polkinghorne, John, 67, 73–74, 149,
  153, 159, 169–70, 177

purpose, 8–14, 26, 87, 99, 105,
  113–15, 149, 157

relativism, 77, 137–38

relativity, 4, 8, 15–19, 29, 36, 44, 70,
  77, 166

relevance, 5–8, 13–14; mutual, 3–19,
  22, 26, 28, 122, 142, 164–66

religion, 7, 116, 120, 155–56, 164

Ruse, Michael, 125–28, 134–39, 158,
  170–71, 177

Russell, Robert, 15, 39, 49–50, 72–73,
  113, 177

secularity, x–xi, 3–4, 12, 28, 69–71,
  137–43, 162

security, 3, 12, 167

sociobiology, 115–42, 150–53, 158

subuniverse, 33, 36–46, 57, 66–87,
  90, 97–100, 104, 113–16, 155–60

suffering, 6, 29, 68, 71, 147–59

survival, 128–43, 150, 158–59

technology, 3, 6, 13, 59–64, 139, 152

time, 7, 15–17, 21, 25, 28, 33–41, 57,
  66, 73, 82, 84–85, 100, 106,
  112–13, 143, 157, 160, 162, 167

truth, 3, 60–61, 77–78, 89, 118, 132,
  135, 140–41, 154–56, 161, 164, 167

universe-as-a-whole (universe), 51–58,
  61–68, 75, 81, 103

values, 6, 13, 39, 44, 59, 99–106,
  114, 117, 120–21, 126–32, 135–41

wholeness, 14–30, 35, 37, 50–56,
  64–65, 73–75, 81–90, 97–109,
  112–14, 122, 129, 133, 158–61,
  167

wholeness-with-diversity, 14, 16, 23,
  26, 81, 84, 101–14, 125, 152

will 130; divine, 48, 154–55; divine
  free, 42, 95, 98–100; free, 50,
  63–74, 95–103, 149–50; moral,
  140–41

Wilson, Edward, 116, 120, 130–42,
  170–71, 174, 177–78

wisdom, 103, 140–42, 161, 167